Interdisciplinary Concepts
in Cardiovascular Health

Ichiro Wakabayashi • Klaus Groschner
Editors

Interdisciplinary Concepts in Cardiovascular Health

Volume II: Secondary Risk Factors

Editors
Ichiro Wakabayashi
Department of Environmental
and Preventive Medicine
Hyogo College of Medicine
Hyogo
Japan

Klaus Groschner
Center for Physiological Medicine
Institute of Biophysics
Medical University of Graz
Graz
Austria

ISBN 978-3-319-01049-6   ISBN 978-3-319-01050-2 (eBook)
DOI 10.1007/978-3-319-01050-2
Springer Heidelberg New York Dordrecht London

Library of Congress Control Number: 2013946028

© Springer International Publishing Switzerland 2013
This work is subject to copyright. All rights are reserved by the Publisher, whether the whole or part of the material is concerned, specifically the rights of translation, reprinting, reuse of illustrations, recitation, broadcasting, reproduction on microfilms or in any other physical way, and transmission or information storage and retrieval, electronic adaptation, computer software, or by similar or dissimilar methodology now known or hereafter developed. Exempted from this legal reservation are brief excerpts in connection with reviews or scholarly analysis or material supplied specifically for the purpose of being entered and executed on a computer system, for exclusive use by the purchaser of the work. Duplication of this publication or parts thereof is permitted only under the provisions of the Copyright Law of the Publisher's location, in its current version, and permission for use must always be obtained from Springer. Permissions for use may be obtained through RightsLink at the Copyright Clearance Center. Violations are liable to prosecution under the respective Copyright Law.

The use of general descriptive names, registered names, trademarks, service marks, etc. in this publication does not imply, even in the absence of a specific statement, that such names are exempt from the relevant protective laws and regulations and therefore free for general use.

While the advice and information in this book are believed to be true and accurate at the date of publication, neither the authors nor the editors nor the publisher can accept any legal responsibility for any errors or omissions that may be made. The publisher makes no warranty, express or implied, with respect to the material contained herein.

Printed on acid-free paper

Springer is part of Springer Science+Business Media (www.springer.com)

Preface

According to the recent WHO statistics, cardiovascular diseases are the leading cause of death globally with an estimated 17.3 million people dying from cardiovascular disease in 2008, representing 30 % of all global deaths. Of these deaths, an estimated 7.3 and 6.2 million were attributed to coronary heart disease and stroke, respectively. Undisputedly, cardiovascular diseases are predicted to remain as the leading cause of death. Over the past decades, we have gained substantial knowledge about the molecular and cellular mechanisms underlying disease initiation and progression and in parallel have identified a plethora of potential risk factors. Moreover, epidemiology has recently moved closer to molecular pathology, enabling a first glimpse to be captured of essential molecular mechanisms that determine cardiovascular risk within a certain population. This book is an attempt to provide an overview of the recent development in this field.

There is common consent that pathogenesis of cardiovascular disorders is mainly based on atherosclerosis, which progresses with age. Thus, retardation of atherosclerosis progression is considered the most effective strategy to prevent cardiovascular morbidity and mortality. A wide range of risk factors are involved in the pathogenesis of atherosclerosis. Lifestyle-related factors including diet, nutrition, physical activity, habitual smoking and alcohol consumption, socioeconomic factors, and psychological stress as modifiable factors, in addition to age, gender, race/ethnicity, and genetic polymorphisms as destined factors, are recognized to determine the risk for cardiovascular events. These "primary risk factors" typically promote the genesis of disorders that may be understood as "secondary risk factors" such as obesity, hypertension, diabetes, dyslipidemia, hyperuricemia, and metabolic syndrome. Thromboatherosclerotic alterations in arterial wall structure result from a combination of secondary risk factors and lead to terminal cardiovascular events such as ischemic heart disease and stroke. Our book is accordingly structured into sections that give detailed information about our current understanding of primary and secondary risk factors, as well as on terminal cardiovascular events. To sustain cardiovascular health, prevention or compensation of primary risk determinants, as well as early diagnosis and treatment of secondary risk factors, are undoubtedly key strategies.

Accumulated groundbreaking insights into cellular mechanisms involved in the pathogenesis of cardiovascular aging and disease include the identification of vasoactive prostanoids, lipoprotein receptors, and nitric oxide. The general scientific value of these basic findings, as well as their wider impact on our society, is well documented by the Nobel Prizes given for each of these discoveries. This book aims to introduce established principles of both cardiovascular epidemiology and molecular pathophysiology and makes a unique attempt to bridge the gap between epidemiological knowledge and current molecular concepts in cellular pathophysiology. The authors spotlight future avenues for research in basic pathophysiology, as well as in cardiovascular therapy and prevention. The comprehensive overview of cardiovascular pathophysiology provided with this book is expected to help readers to address questions on unresolved pathomechanisms and/or to interpret novel epidemiological findings on cardiovascular disorders.

Finally, the editors would like to express their sincere appreciation to all contributors for their dedicated collaboration in this project. We wish to additionally thank Ms. Karin Osibow for her competent and patient support in editing this book.

We hope our book will enable readers to connect epidemiological knowledge with principles of molecular pathophysiology, thereby promoting the development of new strategies for sustaining cardiovascular health.

Hyogo, Japan Ichiro Wakabayashi
Graz, Austria Klaus Groschner

Contents

Volume 2: Secondary Risk Factors

1 **Hypertension and Vascular Dysfunction** 1
   Raouf A. Khalil

2 **Obesity and Diabetes** 39
   Maria Angela Guzzardi and Patricia Iozzo

3 **Hyper- and Dyslipoproteinemias** 63
   Karam M. Kostner and Gert M. Kostner

4 **Hyperuricemia** 87
   Tetsuya Yamamoto, Masafumi Kurajoh, and Hidenori Koyama

5 **Proteomics Toward Biomarkers Discovery
   and Risk Assessment** 115
   Gloria Alvarez-Llamas, Fernando de la Cuesta,
   and Maria G. Barderas

Index 131

Volume 1: Primary Risk Factors

1 **Basic Principles of Molecular Pathophysiology
   and Etiology of Cardiovascular Disorders**
   Michael Poteser, Klaus Groschner, and Ichiro Wakabayashi

2 **Aging**
   Toshio Hayashi

3 **Gender**
   Toshio Hayashi

4 **Gene Polymorphisms and Signaling Defects**
   Christine Mannhalter, Michael Poteser, and Klaus Groschner

5 **Diet and Nutrition to Prevent and Treat Cardiovascular Diseases**
   Hiroshi Masuda

6   **Physical Activity and Cardiovascular Diseases Epidemiology and Primary Preventive and Therapeutic Targets**
    Martin Burtscher and Erich Gnaiger

7   **Alcohol**
    Ichiro Wakabayashi

8   **Tobacco**
    Vicki Myers, Laura Rosen, and Yariv Gerber

9   **Socioeconomic Aspects of Cardiovascular Health**
    Vicki Myers and Yariv Gerber

10 **Stress and Psychological Factors**
    Stefan Höfer, Nicole Pfaffenberger, and Martin Kopp

Index

Volume 3: Cardiovascular Events

1   **Ischemic Heart Disease**
    Yasuhiko Sakata and Hiroaki Shimokawa

2   **Stroke**
    Kazuo Kitagawa

3   **Heart Failure: Management and Prevention of Heart Failure Based on Current Understanding of Pathophysiological Mechanism**
    Masao Endoh

4   **Cardiac Arrhythmias**
    Frank R. Heinzel and Burkert M. Pieske

5   **Pulmonary Hypertension**
    Horst Olschewski and Andrea Olschewski

6   **Peripheral Artery Disease**
    Yoko Sotoda and Ichiro Wakabayashi

7   **Venous Thromboembolism**
    Thomas Gary and Marianne Brodmann

8   **Abdominal Aortic Aneurysm**
    Yasuhisa Shimazaki and Hideki Ueda

Index

Abbreviations

$\cdot O_2^-$	Superoxide anion radical
AC	Adenylate cyclase
ACAT	Acyl-CoA cholesterol acyltransferase
ACE	Angiotensin-converting enzyme
ACh	Acetylcholine
ACS	Acute coronary syndrome
AD	Alzheimer's disease
ADH	Alcohol dehydrogenase
ADMA	Asymmetric dimethyl arginine
ADP	Adenosine diphosphate
AF	Atrial fibrillation
AGE	Advanced glycation end products
Akt/PKB	Akt kinase/protein kinase B
ALDH	Aldehyde dehydrogenase
AMI	Acute myocardial infarction
Ang I/II	Angiotensin I/II
AP-1	Activator protein-1
$AT_{1/2}R$	Angiotensin II receptor type 1, 2
ATP	Adenosine triphosphate
AUC	Area under the curve
AVP	Arginine vasopressin
BMI	Body mass index
BNP	Brain/B-type natriuretic peptide
BP	Blood pressure
CAD	Coronary artery disease
CAP	Calponin
CE	Cholesteryl ester
CETP	Cholesteryl ester exchange/transfer protein
cGMP	Cyclic guanosine monophosphate
CHD	Coronary heart disease
CHF	Chronic heart failure
CIMT	Carotid artery intima–media thickness
CITP	Collagen type I

CKD	Chronic kidney disease
CO	Cardiac output
COX-1, COX-2	Cyclooxygenase-1, cyclooxygenase-2
CRP	C-reactive protein
CSF	Cerebrospinal fluid
CV	Cardiovascular
CVD	Cardiovascular disease
D&HLP	Dys- and hyper-/hypolipoproteinemias
DAG	Diacylglycerol
DGAT1	Diacylglycerol acetyltransferase 1
DM	Diabetes mellitus
DOCA	Deoxycorticosterone acetate
EAT	Epicardial adipose tissue
EC	Endothelial cell
ECE	Endothelin-converting enzyme
ECM	Extracellular matrix
EDCF	Endothelium-derived contracting factor
EDHF	Endothelium-derived hyperpolarizing factor
EDL	Endothelial lipase
EDRF	Endothelium-derived relaxing factor
eGFR	Estimated glomerular filtration rate
EGFR	Epidermal growth factor receptor
eNOS	Endothelial nitric oxide synthase
ER	Endoplasmic reticulum
ERK	Extracellular signal-regulated kinase
ET-1	Endothelin 1
$ET_{A/B}R$	Endothelin receptor A, B
FATP	Fatty acid transport protein
FC	Free cholesterol
FDB	Familial defective-β-lipoproteinemia
FFA	Free fatty acid
FH	Familial hypercholesterolemia
GC	Guanylate cyclase
GC-MS	Gas chromatography on line coupled to mass spectrometry
GFR	Glomerular filtration rate
GLUT4	Glucose transporter type 4
GMP	Guanosine 5′-monophosphate
GWAS	Genome-wide association study
H_2O_2	Hydrogen peroxide
HDL	High-density lipoprotein
HDL-C	High-density lipoprotein cholesterol
HF	Heart failure
HGPRT	Hypoxanthine guanine phosphoribosyl phosphorylase
HL	Hepatic lipase
HMG-CoA	3-hydroxy-methylglutaryl-coenzyme A

hsCRP	High-sensitivity C-reactive protein
HTG	Hypertriglyceridemia
HTN	Hypertension
HUVEC	Human umbilical vein endothelial cell
IDF	International Diabetes Foundation
IDL	Intermediate-density lipoprotein
IL-6	Interleukin-6
IMP	Inosine 5′-monophosphate
IP_3	Inositol 1,4,5-trisphosphate
IRS1	Insulin receptor substrate
IRS1-Ser-P	Insulin receptor substrate – serine phosphate
LCAT	Lecithin–cholesterol acyltransferase
LDL	Low-density lipoprotein
LDL-C	Low-density lipoprotein cholesterol
LDL-R	Low-density lipoprotein receptor
L-NAME	Nitro-L-arginine methyl ester
L-NMMA	N-monomethyl-L-arginine (L-NMMA)
LOX-1	Lectin-like oxidized LDL receptor-1
Lp(a)	Lipoprotein (a)
LPL	Lipoprotein lipase
Lp-PLA$_2$	Lipoprotein-associated phospholipase A$_2$
LP-X	Lipoprotein-X
LV	Left ventricle (LV)
MAP	Mean arterial pressure
MAPK	Mitogen-activated protein kinase
MCP-1	Monocyte chemoattractant protein 1
MEK	MAPK kinase
MEOS	Microsomal ethanol-oxidizing system (MEOS)
MetS	Metabolic Syndrome
MI	Myocardial infarction
MIR	Myocardial insulin resistance
MLC	Myosin light chain
MLCK	Myosin light chain kinase
MLCP	Myosin light chain phosphatase
MMP-1, MMP-2, MMP-9	Matrix metalloproteinase-1, matrix metalloproteinase-2, matrix metalloproteinase-9
MSI	Mass spectrometry imaging
MT1-MMP	MMP activator protein 1
MTP	Microsomal TG-transfer protein
NADH	Nicotin amide adenine dinucleotide phosphate
NADPH	Nicotin amide adenine dinucleotide phosphate, reduced form
NFκB	Nuclear factor kappa-light-chain enhancer of activated B cells
NMR	Nuclear magnetic resonance

nNOS	Neuronal nitric oxide synthase
NO	Nitric oxide
NOS	Nitric oxide synthase
ONOO$^-$	Peroxynitrite
PAD	Peripheral arterial disease
PAI-1	Plasminogen activator inhibitor 1
PDGF	Platelet-derived growth factor
PET	Positron emission tomography
PGI$_2$	Prostaglandin I$_2$, prostacyclin
PGIS	PGI$_2$ synthase
PHLA	Post heparin lipolytic activity
PIP$_2$	Phosphatidylinositol 4,5-bisphosphate
PKC	Protein kinase C
PL	Phospholipid
PLC	Phospholipase C
PLTP	Phospholipid exchange/transfer protein
PMCA	Plasmalemmal Ca^{2+}-ATPase
PPAR-g	Peroxisome proliferator-activated receptor gamma
PPi	Inorganic pyrophosphate
PRPP	5-phosphoribosyl 1-pyrophosphate
RAAS	Renin–angiotensin–aldosterone system
RAGE	Receptors of AGEs
RAS	Renin–angiotensin system
ROC	Receptor-operated Ca^{2+} channel
ROS	Reactive oxygen species
RVR	Renal vascular resistance
SERCA2a	Sarcoplasmic reticulum Ca^{2+}-ATPase 2a
SHR	Spontaneously hypertensive rat
SMCT1	Sodium monocarboxylic transporter 1
SNP	Single nucleotide polymorphism
SOC	Store-operated Ca^{2+} channel
SOD	Superoxide dismutase
sPLA2-IIA$_2$	Secretory type II phospholipase A$_2$
SR	Sarcoplasmic reticulum
SRM	Selected reaction monitoring
STZ	Streptozotocin
T1DM	Type 1 diabetes mellitus
T2DM	Type 2 diabetes mellitus
TD	Tangier disease
TG	Triacylglyceride
TGRLP	Triglyceride-rich lipoprotein
TICE	Trans-intestinal cholesterol excretion
TIMP-1, TIMP-2	Tissue inhibitor of MMP-1, tissue inhibitor of MMP-2
TLR	Toll-like receptor
TNF-α	Tumor necrosis factor alpha

TP	Thromboxane–prostanoid
TPR	Total peripheral resistance
TXA_2	Thromboxane A_2
UCP-1	Uncoupling protein-1
URAT1	Urate transporter 1
VAT	Visceral adipose tissue
VGCC	Voltage-gated Ca^{2+} channel
VLDL	Very low-density lipoprotein
VSM	Vascular smooth muscle
VSMC	Vascular smooth muscle cell
WHO	World Health Organization
WKY	Wistar-Kyoto rat

Hypertension and Vascular Dysfunction

Raouf A. Khalil

Abstract

Blood pressure is tightly regulated by intricate changes in vascular resistance and alterations in cardiac function and the renal control mechanisms of plasma volume. Hypertension is a multifactorial disorder characterized by persistent elevations in blood pressure and involves abnormalities in the cardiac, vascular, and renal function. Increases in heart rate and stroke volume lead to increased cardiac output and could increase blood pressure in young individuals. Vascular endothelial cell dysfunction causes reduction in endothelium-derived relaxing factors, such as nitric oxide, prostacyclin, and hyperpolarizing factor, and increased production of contracting factors such as endothelin-1 and thromboxane A_2. Enhanced vascular smooth muscle contraction mechanisms including $[Ca^{2+}]_i$, protein kinase C, mitogen-activated protein kinase, and Rho kinase could promote vasoconstriction. Also, vascular remodeling and changes in matrix metalloproteinases could affect vascular resistance and blood pressure particularly in aging individuals. Alterations in body fluid regulation by the kidney could cause salt and water retention and increased plasma volume and cardiac output. Activation of the renin-angiotensin system increases angiotensin II in the plasma, leading to generalized vasoconstriction, and locally in the kidney, leading to salt and water retention. Dysregulation of vascular, cardiac, and renal functions could cause persistent increases in blood pressure and hypertension and further increase the risk for progressive cardiovascular disease.

Keywords

Cardiac output • Endothelium • Vascular smooth muscle • Renal function • Arterial pressure

R.A. Khalil, MD, PhD
Division of Vascular Surgery, Harvard Medical School,
Brigham and Women's Hospital, 75 Francis Street, Boston, MA 02115, USA
e-mail: raouf_khalil@hms.harvard.edu

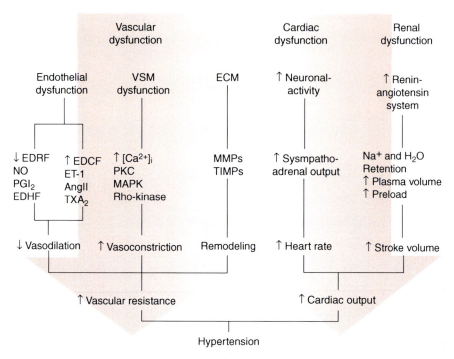

Fig. 1.1 Role of vascular, cardiac, and renal dysfunction in hypertension. The blood vessels through changes in endothelial release of various EDRFs and EDCFs and through changes in signaling mechanisms of vascular smooth muscle contraction could alter the peripheral vascular resistance. Changes in the control mechanisms of the heart and the kidneys alter the cardiac output. Persistent increases in peripheral resistance and cardiac output lead to increases in blood pressure and hypertension. *AngII* Angiotensin II, *NO* nitric oxide, *PGI₂* prostacyclin, *EDHF* endothelium-derived hyperpolarizing factor, *ET-1* endothelin-1, *TXA₂* thromboxane A$_2$, *PKC* protein kinase C, and *MAPK* mitogen-activated protein kinase

1.1 Introduction

Blood pressure (BP) regulation is a critical physiological process involving the combined functions of three systems: the heart, blood vessels, and the kidney. Short-term regulation of BP is achieved by changes in heart rate, stroke volume, and vascular diameter. Long-term regulation of BP may involve vascular remodeling and changes in the renal control mechanisms of plasma volume. An imbalance in the cardiac, vascular, and renal function could cause marked increases in BP, and if these changes are persistent, they could lead to hypertension (HTN) (Fig. 1.1).

HTN is one of the most common and costly diseases. As many as 50 million Americans have high BP, and almost one in every five adults has BP higher than normal (NHANES III). In normal healthy adults, systolic BP is 120–130 mmHg and diastolic BP is 80–85 mmHg. High BP in an adult is defined as systolic BP ≥ 140 mmHg and/or diastolic BP ≥ 90 mmHg. In 90–95 % of cases, the cause of high BP cannot be clearly identified, and hence the term essential HTN. Essential HTN is a multifactorial disorder that involves abnormalities in the heart, blood vessels, and the kidney. Because no single cause of HTN has been identified, different mechanisms may represent the main cause of the disease in different individuals. Heredity and genetic makeup could be a predisposing factor (Spieker et al. 2000). Environmental factors such as dietary sodium intake, obesity, and stress may also play a role in genetically susceptible individuals. This chapter will focus on the role of blood vessels in the control of BP and highlight vascular dysfunction as a major cause of HTN and cardiovascular disease. Because cardiac and renal dysfunctions are important factors in HTN, the role of the heart and the kidney in the control of BP will be briefly described.

1.2 Cardiac Output and Vascular Resistance

The mean arterial pressure (MAP) is determined by the cardiac output (CO) and total peripheral resistance (TPR). MAP=CO×TPR, where CO=stroke volume×heart rate. For comparison between individuals these hemodynamic parameters are corrected for the body surface area (BSA) and presented as the index or "I" parameters such that CI=CO/BSA and TPRI=MAP/CI (Lund-Johansen and Omvik 1990).

Increases in heart rate or force of contraction (inotropy) could increase CO (Fig. 1.1). The increased CO could directly increase BP or trigger cardiogenic reflexes that perpetuate peripheral vasoconstriction. Cardiac remodeling and hypertrophy could also affect the heart's pumping force. In hypertensive patients with a normal left ventricle, the heart may account for 55 % of the increase in systolic BP (Fouad-Tarazi 1988). However, the contribution of abnormalities in the CO vs. TPR varies depending on the severity of HTN. Progression from borderline to moderate and established HTN is associated with changes in cardiac function from a hyperkinetic circulation to normal CI and reduced cardiac function, respectively. In the hyperkinetic circulation stage, hypertensive subjects have higher stroke index, diastolic left ventricular internal dimension index, left ventricular mass index, and wall thickness than normotensive subjects (Lutas et al. 1985). The increased CI in the hyperkinetic type of HTN results from an increase in stroke volume. According to Frank-Starling law, the preload of cardiac muscle before cardiac contraction determines the stroke volume such that a greater end diastolic volume would cause stretching of cardiac muscle fibers and result in greater stroke volume. In a study of young subjects with mild HTN (mean age 31 years, BP 151/95 mmHg) and age-matched normotensive controls, the cardiac muscle sympathetic nerve burst frequency, CO, and CI were markedly higher in hypertensive individuals, and there was no difference in heart rate, TPR, and plasma noradrenaline levels between hypertensive and normotensive subjects (Floras and Hara 1993). Other studies have shown that the characteristic hemodynamic disturbance in young adults (aged 18–40 years) with borderline HTN (readings above and below 140/90 mmHg) or mild HTN (diastolic BP 90–105 mmHg) is high CI and heart rate but "normal" calculated TPR (Frohlich et al. 1970; Julius et al. 1971). However, exercise studies in relatively young subjects with mild HTN show that TPR does not fall as in normotensive age-matched controls. Also, during exercise the heart pump function shows slight reduction with subnormal stroke volume index and CI (Lund-Johansen 1987).

While subjects with borderline or mild HTN may show higher cardiac indexes, some studies suggest that the increases in BP are mainly due to increases in TPR (Lund-Johansen 1986). For instance, borderline hypertensive subjects may have similar CI, but higher heart rate and BP when compared to normotensive controls (Sung et al. 1993). Also, in patients with moderate HTN, an increase in TPR but normal cardiac indices are often observed. As the disease progresses to established HTN, hypertensive subjects may have a pronounced increase in TPR but reduced stroke volume and CO (Fagard 1992), particularly during exercise (Lund-Johansen 1993).

The shift in the hemodynamic pattern of HTN may also vary depending on the subject's age. With age, there is a decrease in CO and maximal heart rate, with an inverse relationship between CO and BP and between CO and age (Fagard 1992). The hemodynamic abnormalities in patients with hyperkinetic borderline HTN change with age. Initially there are "normal" vascular resistance and higher CO, heart rate, and sympathetic tone, but as the disease progresses to moderate HTN, TPR increases and CO decreases to a "normal" level, suggesting adaptation of blood vessels and the heart with age (Amerena and Julius 1995). Also, in established HTN, the hemodynamic pattern changes toward the low CO/high TPR pattern indicating reduced left ventricular compliance and left ventricular hypertrophy (Lund-Johansen 1987). Thus, while in young hypertensives there are signs of increased sympathetic drive and higher-than-normal heart rate and CO, with established HTN further elevation of TPR occurs, while CO drops to subnormal levels particularly during exercise (Lund-Johansen 1987). This is supported by the observation that initial resting CI, heart rate, oxygen consumption, and MAP were 15 % higher in hypertensive patients aged 17–29 years than in normotensive subjects, and the TPRI was similar between both groups. During exercise, lower stroke index and higher TPRI were seen in the hypertensive group. This initially high CI/normal TPRI pattern switched to a low CI/high TPRI pattern at 10- and 20-year follow-up (Lund-Johansen 1991). Also, in young 17–29-year-old vs. 60–69-year-old hypertensive patients with the same MAP, the hemodynamic mechanisms underlying the increased BP are different (Lund-Johansen 1991), with younger patients having an increase in CO and normal TPR, while elderly patients have decreased CO and increased TPR (Messerli et al. 1983). Also, the heart rate, stroke volume, and CI are lower while the TPRI is higher in elderly hypertensives than in young hypertensives. Increased left ventricular posterior wall and septal thicknesses and increased left ventricular mass may also occur in elderly hypertensive subjects. Thus, young hypertensive subjects in the early phase of HTN typically have a high resting CO, whereas in the older hypertensive subjects, CO is low and TPR is high (Lund-Johansen 1991).

1.3 Vascular Control Mechanisms of BP

Blood vessels particularly small resistance arteries with a diameter of ≤200 μm play an important role in the regulation of vascular resistance and BP, and dysregulation of vascular tone could play a role in the increased TPR and BP in HTN (Luscher et al. 1992b). The arterial wall diameter is intrinsically regulated by the vascular endothelial cells (ECs), vascular smooth muscle (VSM), and extracellular matrix (ECM) (Fig. 1.1).

1.3.1 Endothelial Cells (ECs)

ECs are located between VSMCs of the vessel wall and the circulating platelets and blood cells. Such a strategic location allows the endothelium to regulate the function of both VSM and circulating blood cells. ECs can regulate vascular tone by releasing endothelium-derived relaxing factors (EDRFs) including nitric oxide (NO), prostacyclin (PGI_2), and hyperpolarizing factor (EDHF) (Ignarro et al. 1987; Furchgott and Vanhoutte 1989; Busse et al. 2002; Feletou and Vanhoutte 2009) as well as contracting factors (EDCFs) such as endothelin (ET-1), angiotensin II (AngII), thromboxane (TXA_2), superoxide anion ($•O_2^-$), and endoperoxides (Luscher 1990; Vanhoutte 1992) (Fig. 1.1). ECs can also regulate VSMC growth by maintaining a balance between growth promoters and growth inhibitors (Puddu et al. 2000).

1.3.1.1 Nitric Oxide (NO)

NO is the main EDRF and a major vasodilator and regulator of vascular function and BP (Ignarro et al. 1987; Baylis and Qiu 1996; Fleming and Busse 1999). NO is produced by NO synthase (NOS) using L-arginine as substrate. Three NOS isoforms, NOS1 (nNOS), NOS2 (iNOS), and NOS3 (eNOS), have been described (Baylis and Qiu 1996). Activation of ECs increases intracellular Ca^{2+} which in turn activates endothelial eNOS, transforms L-arginine to L-citrulline, and increases NO production (Li et al. 1999). The NO released from ECs diffuses into VSM where it activates guanylate cyclase and increases intracellular cyclic GMP and thereby causes vascular relaxation (Gruetter et al. 1981; Ignarro and Kadowitz 1985) (Fig. 1.2).

EC dysfunction and decreased bioavailability of NO is a major cause of impaired endothelium-dependent relaxation and increased vascular tone in HTN (Luscher 1992; Panza et al. 1995; Vanhoutte and Boulanger 1995; Schiffrin 1996; Taddei et al. 2000a). The availability of L-arginine substrate is an important step in NO production. Asymmetric dimethylarginine (ADMA) is a competitive endogenous inhibitor of NOS. ADMA circulates in concentrations sufficient to cause vasoconstriction that is reversed by L-arginine and is degraded by dimethylarginine dimethylaminohydrolase. Plasma levels of ADMA are increased in hypercholesterolemic rabbits and in patients with hypercholesterolemia and may

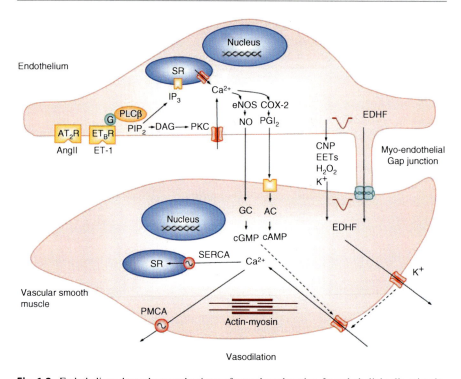

Fig. 1.2 Endothelium-dependent mechanisms of vascular relaxation. In endothelial cells, stimulation of vasodilator receptors such as AT2R and ETBR activates phospholipase C-β (*PLCβ*) and increases the hydrolysis of phosphatidylinositol 4,5-bisphosphate (*PIP$_2$*) into inositol 1,4,5-trisphosphate (*IP$_3$*) and diacylglycerol (*DAG*). IP$_3$ induces Ca^{2+} release from the endoplasmic reticulum (ER). The increased [Ca^{2+}]$_i$ activates endothelial nitric oxide synthase (*eNOS*) and increases nitric oxide (*NO*) production. NO diffuses into vascular smooth muscle (VSM), where it stimulates guanylate cyclase (*GC*) and increases cGMP. cGMP causes vascular relaxation by decreasing Ca^{2+} entry through Ca^{2+} channels, stimulating Ca^{2+} removal via plasmalemmal (*PMCA*) and sarcoplasmic reticulum Ca^{2+}-ATPase (*SERCA*), and decreasing the actin-myosin myofilament force sensitivity to Ca^{2+}. Vasodilator receptors are also coupled to stimulation of cyclooxygenases (*COX-2*) and increased prostacyclin (*PGI$_2$*). PGI$_2$ activates adenylate cyclase (*AC*) and increases cAMP, which causes VSM relaxation by mechanisms similar to those of cGMP. Activation of endothelial receptors also increases the release of endothelium-derived hyperpolarizing factor (*EDHF*), which activates K$^+$ channels and causes VSM hyperpolarization and relaxation. *Dashed arrows* indicate inhibition

be increased in hypertensive patients and contribute to reduced endothelium-dependent vasodilation (Cardillo et al. 1998). However, increased L-arginine availability does not modify endothelium-mediated vasodilation in hypertensive individuals, suggesting that HTN may not be due to deficiency of the NO precursor but rather to defective utilization by NOS (Panza et al. 1993a; Noll et al. 1997). Decreased bioavailability of NO could involve reduced synthesis/release, increased inactivation by reactive oxygen species, and impaired responsiveness of VSM to NO.

Decreased synthesis/release of NO is a major cause of decreased endothelium-dependent relaxation and increased vascular resistance in HTN (Panza et al. 1993b; Panza 1997). In HTN, the decrease in endothelium-dependent relaxation is often associated with impaired activity of constitutive eNOS (Chou et al. 1998). Measurement of forearm blood flow in hypertensive subjects have shown blunted vasodilation to endothelium-dependent agonists such as acetylcholine (ACh) and bradykinin, supporting EC dysfunction and abnormal NO synthesis in HTN (Cardillo et al. 1998). However, endothelium-independent relaxation to sodium nitroprusside is not different in blood vessels of hypertensive and normotensive subjects (Falloon and Heagerty 1994), supporting that the impaired vascular relaxation in hypertensive patients is not due to impaired VSM responsiveness to NO. Impaired endothelial NO production is also observed in offspring of patients with essential HTN (McAllister et al. 1999).

ACh-induced relaxation is reduced, and NO production is diminished in blood vessels of animal models of HTN (Luscher et al. 1992a; Brovkovych et al. 1999). In support of a role of eNOS in BP control, eNOS KO mice have HTN (Huang et al. 1995). Also, in experimental animals, both acute and chronic inhibition of NOS with L-arginine analogs such as Nω-monomethyl-L-arginine (L-NMMA) and Nω-nitro-L-arginine methyl ester (L-NAME) cause an increase in BP. The HTN caused by chronic treatment with NOS inhibitors is associated with cardiac hypertrophy and renal insufficiency and is reversed by co-administration of L-arginine providing further evidence that NO regulates vascular tone (Hedner and Sun 1997).

Oxidative stress could alter EC function and endothelium-dependent modulation of vascular tone in HTN (Taddei et al. 1998; Spieker et al. 2000). Under physiological conditions •O_2^- is removed by superoxide dismutase (SOD). HTN is associated with increased production of •O_2^- and other reactive oxygen species leading to inactivation of NO and increased vascular tone (Cai and Harrison 2000). Decreased SOD activity and increased •O_2^- production would cause NO inactivation and may contribute to the pathogenesis of HTN (Falloon and Heagerty 1994; Jun et al. 1996). Endothelial cell dysfunction may also involve increased arginase activity, which converts L-arginine to L-ornithine and urea thus decreasing the substrate for NO production (Johnson et al. 2005). H_2O_2 upregulates arginase activity and may impair endothelium-dependent NO-mediated dilation of coronary arterioles (Thengchaisri et al. 2006).

1.3.1.2 Prostacyclin (PGI$_2$)

Prostanoids may play a role in the regulation of vascular function and BP. Prostanoids are derived from arachidonic acid by the sequential actions of phospholipase A2, cyclooxygenase (COX), and specific prostaglandin synthases. There are two major COX enzymes, constitutive COX1 and inducible COX2. PGI$_2$ is a major prostanoid produced by ECs and influences many cardiovascular processes. PGI$_2$ diffuses through the intima and acts mainly on IP receptor in VSM but may act on other prostanoid receptors with variable affinities. PGI$_2$/IP interaction triggers G protein-coupled activation of adenylate cyclase and increases cyclic AMP/

protein kinase A in VSM, resulting in decreased $[Ca^{2+}]_i$ and vascular relaxation (Majed and Khalil 2012) (Fig. 1.2).

The contribution of PGI_2 as compared to other vasodilators to vascular relaxation generally depends on the size of the blood vessel, the specific vascular bed, and the animal species. When compared to NO, PGI_2 contributes very little to endothelium-dependent relaxation of rat aorta and large proximal branches of mesenteric arteries (Shimokawa et al. 1996) and human proximal gastroepiploic arteries (Urakami-Harasawa et al. 1997). In contrast, EDHF is the major EDRF in small resistance rat mesenteric arteries and human gastroepiploic arteries that are known to control vascular tone and BP. Also, *in vivo* data have shown that endogenous PGI_2 may have little role in the regulation of BP when compared to NO because pharmacological blockade of COX with indomethacin or aspirin does not substantially affect BP in humans or rats, while pharmacological blockade or genetic ablation of NOS has a profound effect on BP, increasing it by 30–50 mmHg. Although some studies have shown that mice lacking PGI_2 synthase (PGIS) have greater BP than wild type, this has not been attributed to changes in PGI_2-mediated relaxation but rather to the associated gross abnormalities in the kidney or the reduction in the lumen-to-wall ratio in the aorta and pulmonary and renal vessels caused by the decreased PGI_2 synthesis (Parkington et al. 2004).

Decreased PGI_2 production and abnormally high TXA_2/PGI_2 ratio have been implicated in the pathogenesis of pulmonary arterial HTN, an often fatal disease (Majed and Khalil 2012). Also, while PGI_2 is largely known as a vasodilator (Miller 2006), some *ex vivo* studies suggest that, depending on the vessel type and/or concentration tested, PGI_2 may induce VSM contraction (Williams et al. 1994). In rat aorta precontracted with norepinephrine, PGI_2 elicits a biphasic response such that lower concentrations elicit relaxation, while at higher concentrations the relaxation decreases (Williams et al. 1994). High PGI_2 concentrations also cause contraction in cat (Uski et al. 1983), dog (Chapleau and White 1979), monkey (Kawai and Ohhashi 1994), and human cerebral arteries (Uski et al. 1983). The endothelium may play a role in mediating the vasoconstrictor effect of PGI_2 whereby PGI_2-induced endothelial NO release through IP receptor is counterbalanced by PGI_2-induced activation of endothelial EP1 and thromboxane-prostanoid (TP) receptor, which could cause IP receptor desensitization and internalization, leading to reduced NO release and decreased vascular relaxation (Xavier et al. 2009). In aortic rings of spontaneously hypertensive rat (SHR) and Wistar-Kyoto rat (WKY), PGI_2 unexpectedly acts as an EDCF, likely due to decreased IP receptor expression and PGI_2-mediated activation of TP receptor (Gluais et al. 2005). The paradoxical contractile effects of PGI_2 could also be related to the amount of PGIS expressed in vascular tissues. PGIS gene expression in ECs appears to be augmented with aging and in HTN. It has been shown that during endothelium-dependent contraction of SHR aorta to ACh, PGI_2 production is larger than that of other prostanoids and reaches levels compatible with activation of TP receptors in VSM (Vanhoutte 2011). In these rat models, inhibition of PGIS by tranylcypromine further enhances ACh-induced endothelium-dependent contraction likely due to enhanced PGH_2 spillover, a more potent TP receptor agonist than PGI_2 (Gluais et al. 2005). The vasoconstrictor effects of PGI_2 may increase the cardiovascular risk, and further investigation of the mechanisms involved may identify new therapeutic targets in vascular disease (Majed and Khalil 2012).

1.3.1.3 EDHF

ECs may influence the vascular tone by releasing EDHF. EDHF hyperpolarizes ECs by activating K^+ channels. Hyperpolarization is transferred to VSMCs via myoendothelial gap junctions (connexins), or the efflux of K^+ ions from hyperpolarized ECs may accumulate in the extracellular space and act on the Kir and Na^+/K^+ pump to cause VSM hyperpolarization. The nature and mechanism of action of EDHF vary in different tissues and species. In addition to K^+, other putative EDHFs include EETs, H_2O_2, and C-type natriuretic peptide (Feletou and Vanhoutte 2006) (Fig. 1.2). Administration of 40Gap 27, an inhibitory peptide homologous to the second extracellular loop of connexin 40, abolishes the residual ACh vasodilatation, supporting a role for gap junctions in EDHF-mediated vasodilation. Also, administration of 40Gap 27, even in the absence of L-NAME and indomethacin, is associated with decreased renal blood flow and increased BP, suggesting that tonic EDHF may maintain basal renal vascular tone and blood flow (De Vriese et al. 2002). Connexin 40-deficient mice are hypertensive, suggesting that impaired EDHF-mediated vasodilatation may play a role in HTN (de Wit et al. 2000). Also, while EDHF-mediated renal vascular relaxation is normal in young SHR, it is reduced in aged SHR and associated with depolarized resting membrane potential of the renal artery VSM (Bussemaker et al. 2003), suggesting that decreased EDHF may be involved in the progression of HTN.

1.3.1.4 EETs and HETEs

Epoxy and 20-hydroxy derivatives of arachidonic acid oxidation by cytochrome P450 monooxygenase may play a role in the regulation of vascular function, BP, and the pathogenesis of HTN. EETs are produced by ECs, renal proximal tubules, and collecting ducts, and some of the EETs are potent vasodilators and natriuretic compounds.

20-HETE may play a role in renal myogenic autoregulation and vasoconstrictor actions of AngII and ET-1 (Ma et al. 1993; Alonso-Galicia et al. 2002). In addition to its potent renal vasoconstrictor effects, 20-HETE is also natriuretic and therefore has counteracting effects on BP. The vasoconstrictor action of 20-HETE involves blockade of Ca^{2+}-activated K^+ channels in VSM. Renal expression of CYP4A2 which produces 20-HETE is increased in SHR, and CYP4A2 inhibitors attenuate HTN in SHR and deoxycorticosterone acetate (DOCA) salt-sensitive HTN (Schwartzman et al. 1996). Also, in mice, deletion of CYP4A14 increases expression of CYP4A2, resulting in increased 20-HETE and HTN (Imig 2005).

Hydroxy derivatives of arachidonic acid 12- and 15-HETE are produced by 12/15-lipooxygenase in the glomeruli, mesangial cells, and renal microvascular ECs and VSMCs (Zhao and Funk 2004). 12- and 15-HETE constrict renal vessels and glomerular mesangial cells. 12-HETE induces VSMC growth and partly mediates AngII-induced afferent arteriolar vasoconstriction as well as TGF-β and

AngII-induced mesangial cell hypertrophy and ECM accumulation (Yiu et al. 2003). 12/15-lipooxygenase inhibition or gene deletion blunts the pressor response to AngII and blocks AngII-induced hypertrophy and ECM accumulation in rat mesangial cells (Kim et al. 2005). Also, 12/15-lipooxygenase deficiency is associated with increased eNOS expression/activity, suggesting a relation between the two pathways in the control of vascular function and BP (Anning et al. 2005). Urinary excretion of 12-HETE is increased in HTN. Also, 12-HETE production is increased in SHR, and 12-lipooxygenase inhibitors decrease BP in these rats (Sasaki et al. 1997).

1.3.1.5 Endothelin (ET-1)

Endothelial dysfunction may be associated with increased production of EDCFs such as ET-1 (Luscher 1992; Panza et al. 1995; Taddei et al. 1996; Vanhoutte 1996; Hedner and Sun 1997). The ET-1 produced by ECs acts mainly in a paracrine fashion on underlying VSMCs to cause vasoconstriction (Yanagisawa et al. 1988; Pollock et al. 1995; Schiffrin and Touyz 1998) (Fig. 1.3). In addition to its potent vasoconstrictor effect, ET-1 induces VSM growth and vascular hypertrophy (Spieker et al. 2000). Increased ET-1 synthesis may account for many of the features of HTN, including increased vascular tone and vascular hypertrophy (Nava and Luscher 1995; Pollock et al. 1995; Schiffrin and Touyz 1998; Goddard and Webb 2000). Also, the prominent and long-acting vasoconstrictor effects of ET-1 have suggested that it may play a role in the regulation of BP and the pathogenesis of HTN. For example, in the peripheral circulation of normotensive subjects, where tonic NO production is preserved, blockade of ET receptors by the nonselective ETA/B receptor antagonist TAK-044 causes modest vasodilation. In contrast, in hypertensive subjects, where NO production is reduced, the vasodilator effects of TAK-044 are more evident, indicating prominent vasoconstrictor effects of endogenous ET-1 (Taddei et al. 2000a).

However, a clear relationship between ET-1 and HTN has not been established. For instance, the circulating levels of ET-1 are not increased in all animal models of HTN and in human HTN (Luscher 1992). However, localized vascular production of ET-1 and its levels in the vicinity of VSMCs may still be increased (De Artinano and Gonzalez 1999). Also, blood vessels of hypertensive animals may show increased responsiveness to ET-1 (Nava and Luscher 1995; De Artinano and Gonzalez 1999), and ET-1 may potentiate the vasoconstrictor effects of noradrenaline and AngII in HTN (Luscher 1992). Most patients with HTN show normal or slightly increased ET-1 levels (Schiffrin 1995). Among African Americans, plasma ET-1 levels are increased in hypertensive compared with normotensive controls. On the other hand, among individuals with similar severity of HTN, plasma ET levels are not higher in African Americans compared with Caucasians (Schiffrin 2001). Upregulation of the ET system is often observed in severe cases of HTN associated with coronary artery disease, heart failure, atherosclerosis, and pulmonary HTN. For example, plasma levels of ET-1 (5.15 pg/mL) and big ET-1 (25.7 pg/mL) are markedly elevated in patients with heart failure compared with control subjects (0.75 and 7.7 pg/mL, respectively) (Modesti et al. 2000). The 24-h urinary excretion

1 Hypertension and Vascular Dysfunction

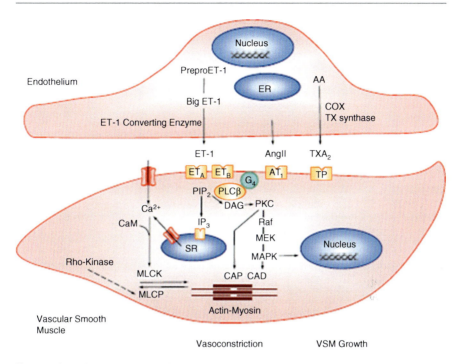

Fig. 1.3 Signaling mechanisms of vascular smooth muscle contraction. In endothelial cells, pre-proET-1 is cleaved into big ET-1 which is cleaved by endothelin-converting enzyme (ECE) into active ET-1. ET-1 is released from ECs and acts in a paracrine fashion on VSM. Other EDCFs include AngII and TXA$_2$. In VSM, the interaction of ET-1 with ET$_A$R and ET$_B$R, AngII with AT$_1$R, and TXA$_2$ with TP receptors activates PLCβ and increases IP$_3$ and DAG. IP$_3$ stimulates Ca^{2+} release from the sarcoplasmic reticulum (*SR*). Vasoconstrictor agonists also stimulate Ca^{2+} entry through Ca^{2+} channels. Ca^{2+} binds calmodulin to form a complex, which causes activation of myosin light chain (*MLC*) kinase (*MLCK*), MLC phosphorylation, actin-myosin interaction, and VSM contraction. DAG activates PKC, which phosphorylates the actin-binding protein calponin (*CAP*) or initiates a protein kinase cascade involving Raf, MAPK kinase (*MEK*), and MAPK, leading to phosphorylation of caldesmon (*CAD*) and thereby increases the myofilament force sensitivity to Ca^{2+}. Agonists could also activate Rho kinase, which inhibits MLC phosphatase (*MLCP*) and increases MLC phosphorylation. Agonist-mediated activation of MAPK could also induce gene transcription and VSM growth and proliferation. *Dashed arrows* indicate inhibition

of ET-1 is also greater in hypertensive patients with heart failure (17.0 ng/g urinary creatinine or UC) compared to control subjects (1.7 ng/g UC) (Modesti et al. 2000).

The discrepancy in the plasma ET-1 levels among hypertensive patients may be due to its rapid clearance from the blood stream. Also, ET-1 is mainly secreted in a polarized fashion from ECs to the underlying VSM, leading to minimal increases in its circulating plasma levels. For example, ET-1 mRNA expression is increased in the endothelium of subcutaneous resistance arteries from patients with moderate to severe HTN (Schiffrin 1995). ET-1 tissue expression is also increased in salt-sensitive HTN, low-renin HTN, and obesity and insulin resistance-related HTN (Touyz and Schiffrin 2003).

The tissue levels of ET-1 may also show variability in experimental HTN. ET-1 levels are elevated in the aortic wall of DOCA-salt hypertensive rats (730 pg/g) compared to control rats (120 pg/g) (Zhao et al. 2000). Also, the tissue levels of ET-1 are markedly higher than the plasma levels, supporting that ET-1 tissue levels could be a better indicator of the vascular ET-1 system in HTN. ET-1 vascular tissue levels are also increased in DOCA-salt-treated SHR and salt-loaded stroke-prone SHR, Dahl salt-sensitive rats, AngII-infused rats, and 1-kidney 1-clip Goldblatt hypertensive rats, but not in SHR, 2-kidney 1-clip hypertensive rats, or L-NAME-treated rats (Schiffrin 1998).

ET-1 activates ET_AR in VSM to produce vasoconstriction (Fig. 1.3) and ET_BR mainly in ECs to induce the release of relaxing factors and promote vasodilation (Fig. 1.2). Both ET_AR and ET_BR could play a role in the regulation of BP. An increase in the amount/activity of ET_AR or a decrease in the amount/activity of ET_BR is expected to cause an increase in BP and HTN. However, the relationship between ET_AR and ET_BR in HTN could be more complex, as ET_BR is expressed in VSM and could activate VSM contraction mechanisms and produce vasoconstriction (Fig. 1.3).

The number of vascular ET receptors could vary in different forms of HTN and in various tissues isolated from subjects with the same form of HTN. For example, ET_BR is upregulated in the kidneys of DOCA-salt hypertensive rats, consistent with a role for ET_BR in the renal regulation of BP (Pollock and Pollock 2001). ET receptors could also be downregulated by ET-1 when a large amount of ET-1 is produced in the vasculature. For instance, in DOCA-salt hypertensive rats, the ET receptor density is reduced in some vascular beds, possibly due to increased vascular production of ET-1 and consequent downregulation of ET receptors (Nguyen et al. 1992).

The variability in the ET system in HTN may not only involve the plasma and vascular tissue levels of ET and the amount of vascular ET receptors but could also involve the vascular response to ET. ET-1 induced contraction is increased in the coronary arteries of rat hearts during ischemia/reperfusion (de Groot et al. 1998) and in the pulmonary artery of rat models of pulmonary HTN (McCulloch et al. 1998; Barman 2007). The increased vascular reactivity to ET-1 in some forms of experimental HTN may be related to increased $[Ca^{2+}]_i$ in VSM (Tostes et al. 1997; Schroeder et al. 2000). In contrast, ET-induced contraction is not enhanced in the aorta of SHR and is even decreased in mesenteric arteries of DOCA-salt hypertensive rats and in ET_BR-deficient rats possibly due to reduced ET_AR density as a result of its downregulation by increased vascular ET-1 production or decreased ET_BR-mediated ET-1 clearance (Nguyen et al. 1992; Perry et al. 2001).

In milder forms of HTN, the VSM of resistance arteries are restructured around a smaller lumen without true hypertrophy, resulting in reduced circumference and amplification of pressor stimuli. An example of this structurally based amplification is the enhanced vasoconstriction in isolated microvessels of SHR and renovascular hypertensive rats (Schiffrin 1995). In severe forms of HTN and in secondary HTN, hypertrophic remodeling of VSM occurs (Touyz and Schiffrin 2003). Large conduit artery such as the aorta may demonstrate thickened tunica media, increased collagen deposition, and decreased compliance, leading to increased systolic BP and pulse pressure. Because ET-1 is a growth promoter, it could play a role in the VSM hypertrophy observed in severe HTN and in DOCA-salt hypertensive rats (Schiffrin 1995). The responsiveness of endothelial ET_BR may also change in HTN. ET_BR-mediated vasorelaxation is greater in SHR and DOCA-salt hypertensive rats than normotensive rats. Thus, while ET_AR may play a role in the development of DOCA-salt-sensitive induced HTN, ET_BR may protect against vascular and renal injury (Matsumura et al. 1999). ET-1 may decrease the release of EDRFs and thereby further enhance vasoconstriction in blood vessels of SHR (Schiffrin 1995).

1.3.1.6 Angiotensin II (AngII)

AngII is a major mediator of vascular dysfunction and vasoconstriction (Fig. 1.3). AngII stimulates AT_1R in VSM to produce vasoconstriction and AT_2R largely in ECs to induce vasodilation (Fig. 1.2). AT_1R are localized on VSMCs of blood vessels particularly the renal afferent and efferent arterioles. AngII also contributes to regulation of salt and water balance directly by activating the NHE3 transporter and stimulating sodium reabsorption in the proximal tubules and indirectly by stimulating aldosterone secretion in the adrenals (Liu and Cogan 1988; Wang and Chan 1990). Some of the components of the renin-angiotensin system (RAS) are upregulated in certain forms of HTN. For instance, intrarenal angiotensinogen mRNA is increased in Dahl salt-sensitive rats on a high-salt diet (Kobori et al. 2003).

Increased systemic and renal vascular sensitivity to AngII is observed in normotensive men with family history of HTN (Widgren et al. 1992) and in SHR even in the developmental phase of HTN (Chatziantoniou et al. 1990; Vyas and Jackson 1995). Also, sympathetic nerves enhance the renovascular responses to AngII in SHR (Dubinion et al. 2006). Interestingly, low-dose infusion of AngII for 6–10 days in rabbits, rats, or mice causes increases in renal vascular resistance (RVR), BP, oxidative stress, urinary 8-iso-PGF2α, peroxynitrite (ONOO$^-$) generation, and nitrotyrosine deposition. AngII induces ET-1 production, which increases •O$_2^-$ and decreases NO bioavailability. AngII also promotes the renal production of TXA$_2$ which in turn causes renal vasoconstriction (Vagnes et al. 2007). The effects of AngII on multiple systems and vascular mediators may explain the elevated BP in essential and renovascular HTN where the AngII levels may be normal (Romero and Reckelhoff 1999).

The net effects of AngII on the vascular and renal mechanisms and BP may also depend on the duration of increased AngII activity. Short-term infusion of AngII in rats causes an elevation in BP that returns to normal upon cessation of AngII. In contrast, prolonged AngII infusion for 6–10 days causes long-lasting elevation in BP, afferent and efferent arteriolar constriction, and reduced glomerular filtration rate (GFR) that persists after cessation of AngII infusion (Franco et al. 2001). Long-term AngII may cause injury to the renal microvessels, and peritubular capillary and tubulointerstitium leading to structural changes such as afferent arteriolar wall thickening and loss of peritubular capillaries, as well as functional changes such as decreased NO and increased ET-1. The AngII-induced tubulointerstitial injury also promotes the release of pro-inflammatory chemokines and leukocyte adhesion proteins and infiltration of mononuclear cells, which express AngII at the site of injury and thereby further augment intrarenal AngII (Johnson et al. 1992). The increased AngII leads to further renal vasoconstriction, microvascular wall thickening, decreased glomerular filtration, sodium retention, and salt-sensitive HTN (Franco et al. 2007).

1.3.1.7 Thromboxane A$_2$ (TXA$_2$)

TXA$_2$ is another prostanoid produced by COX from arachidonic acid (Fig. 1.3) and an important mediator of the renal hemodynamic and pressor effects of AngII. TXA$_2$ is a potent vasoconstrictor in the renal vessels and afferent arterioles acting via TP receptor to increase renal VSM [Ca^{2+}]$_i$. In isolated microperfused rabbit afferent arterioles, TP receptor activation produces reactive oxygen species, which promote vasoconstriction, but also produces NO which counteracts vasoconstriction (Schnackenberg et al. 2000). An increase in endogenous levels of TXA$_2$ may contribute to the enhanced tubuloglomerular feedback activity in young SHR, leading to a rightward shift in the pressure-natriuresis curve and HTN (Brannstrom and Arendshorst 1999). TP receptor may also mediate the effects of the F2 isoprostane 8-iso-PGF$_{2\alpha}$ on preglomerular vasoconstriction. The renal vascular reactivity to TXA$_2$ is increased in SHR possibly due to increased affinity/number of renal TP receptor (Chatziantoniou et al. 1990).

1.3.1.8 EC Dysfunction and HTN: Cause or Consequence

EC dysfunction has been associated with HTN in humans and in animal models of HTN including SHR and Dahl salt-sensitive rats (Panza et al. 1990; Luscher et al. 1991; Schiffrin 1996; Taddei et al. 2000a). In HTN, there is impaired endothelium-dependent relaxation and decreased production and/or action of NO, PGI$_2$, and EDHF (Luscher 1992; Panza et al. 1995; Taddei et al. 1996; Vanhoutte 1996). The decreased endothelium-dependent relaxation could promote vasoconstriction and contribute to the increased vascular resistance in HTN. Increased production of or sensitivity to ET-1, AngII, and TXA$_2$ particularly in resistance vessels would further increase vasoconstriction and vascular resistance (Nava and Luscher 1995; Pollock et al. 1995; Schiffrin and Touyz 1998; Goddard and Webb 2000) (Fig. 1.1). Some vasoconstrictors are also mitogenic and growth promoters leading to VSM growth and proliferation. An imbalance in EC production of vasodilator and vasoconstrictor substances would lead to further vasoconstriction and vascular and end-organ damage (Luscher and Noll 1995; Puddu et al. 2000; Spieker et al. 2000).

Although EC dysfunction is present in most forms of HTN, whether it is the cause or consequence of HTN is unclear. Studies in normotensive offspring of parents with essential HTN demonstrated endothelial dysfunction, deceased basal NO production, and reduced ACh-mediated vasodilation in the forearm of subjects at high risk of essential HTN (McAllister et al. 1999). The reduced ACh-mediated vasodilation in the forearm was linked to a defect in the NO pathway, and suggested that impaired NO production precedes the onset of HTN (Taddei et al. 2000b). Also, NOS blockade by L-NAME and related agents in rats causes HTN, and L-arginine has antihypertensive effect in salt-loaded Dahl rats, supporting a causative

1 Hypertension and Vascular Dysfunction

association between endothelial dysfunction and HTN (Van Zwieten 1997). Also, adult SHRs have both endothelial dysfunction and overt HTN, while young SHRs have endothelial dysfunction but normal BP, suggesting that endothelial dysfunction may be a cause rather than a consequence of HTN (Jameson et al. 1993).

Endothelial dysfunction may not only initiate the increase in BP but also maintain it. Due to its position between the blood stream and VSMCs, ECs act as both a mediator and a target of HTN (Luscher 1994; Spieker et al. 2000). Studies in hypertensive animal models and in human subjects have demonstrated an association between increased BP and impaired endothelium-dependent vascular relaxation (Cardillo et al. 1998; Cardillo and Panza 1998). Because endothelial dysfunction is observed in various types of HTN and is often reversed by correcting BP, some studies suggested that endothelial dysfunction may not be a primary cause but rather secondary to HTN (Luscher 1994; Naruse et al. 1994; Noll et al. 1997; Kiowski 1999). At later stages of HTN, endothelial dysfunction could cause further increases in vascular resistance and cardiovascular complications (Luscher 1994). HTN could cause a decrease in NO production in ECs, thereby diminishing their vasorelaxant effects. HTN could also increase the expression of adhesion molecules and increases the adherence of leukocytes to the vessel wall. These pathological alterations in the endothelium secondary to HTN could initiate and accelerate the pathogenesis of chronic vascular disease (Haller 1996). In experimental animals, the degree of impairment of endothelium-dependent responses is positively correlated with BP and becomes more pronounced as HTN develops and with increased duration of HTN (Luscher 1994). Thus, a significant portion of endothelial dysfunction may occur as a consequence of high BP, leading to further increases in vascular resistance at later stages of the disease and the incidence of vascular complications such as atherosclerosis (Boulanger 1999).

1.3.2 Vascular Smooth Muscle (VSM)

VSM tone is a major regulator of vascular resistance and BP. Persistent increases in vasoconstriction could lead to increased vascular resistance and HTN. The vascular reactivity to norepinephrine is greater in SHR compared WKY rats (Mulvany and Nyborg 1980). Although the median effective concentration (EC50) for norepinephrine is not different in the two strains, the rate of tension development and the maximal tension are greater in SHR than WKY (Cauvin and van Breemen 1985; Crews et al. 1999), suggesting increases in the signaling mechanisms downstream from receptor activation.

Although VSM can maintain tone for long periods of time with very little energy expenditure, the mechanisms involved are not fully understood. The classic Ca^{2+}-dependent MLC phosphorylation pathway cannot fully explain all modes of VSM contraction, and additional signaling pathways have been suggested to increase the myofilament force sensitivity to Ca^{2+}, including protein kinase C and Rho kinase (Fig. 1.3). VSM dysfunction involves an increase in basal tone and VSM reactivity to vasoconstrictors and increased signaling pathways of VSM contraction.

1.3.2.1 Ca^{2+}-Dependent MLC Phosphorylation

Ca^{2+}-dependent phosphorylation of the 20 kDa MLC is a major determinant of VSM tone (Rembold and Murphy 1988; Kamm and Stull 1989). Stimulation of α-adrenergic receptors by agonists such as norepinephrine or phenylephrine and the interaction of ET-1, AngII, and TXA$_2$ with their corresponding receptors in VSM activate phospholipase C which stimulates the hydrolysis of the plasma membrane phosphatidylinositol 4,5-bisphosphate (PIP$_2$) and increases the formation of inositol 1,4,5-trisphosphate (IP$_3$) (Fig. 1.3). IP$_3$ stimulates Ca^{2+} release from the intracellular stores in the sarcoplasmic reticulum causing an initial increase in [Ca^{2+}]$_i$. The agonist also stimulates Ca^{2+} entry from the extracellular space through receptor-operated (ROCs), store-operated (SOCs), and voltage-gated Ca^{2+} channels (VGCCs) causing maintained increase in [Ca^{2+}]$_i$. Ca^{2+} binds calmodulin to form a Ca^{2+}-calmodulin complex, which stimulates MLC kinase, increases MLC phosphorylation, and initiates actin-myosin interaction and VSM contraction (Fig. 1.3). VSM relaxation is initiated by a decrease in [Ca^{2+}]$_i$ due to uptake by the sarcoplasmic reticulum and extrusion by the plasmalemmal Ca^{2+} pump and the Na$^+$/Ca^{2+} exchanger. The decrease in [Ca^{2+}]$_i$ causes dissociation of the Ca^{2+}-calmodulin complex, and the phosphorylated MLC is dephosphorylated by MLC phosphatase.

VSM [Ca^{2+}]$_i$ is a major determinant of the myogenic mechanism of renal vascular autoregulation and agonist-induced vasoconstriction (Roman and Harder 1993; Murphy et al. 2003). The renal microcirculation has heterogeneous distribution of Ca^{2+} channels and differential Ca^{2+} influx in various vascular segments. The renal cortical preglomerular afferent arteriole, juxtamedullary efferent arteriole, and outer medullary vasa recta have L- and T-type VGCCs, whereas the mid-to-outer cortical efferent arteriole doesn't (Hansen et al. 2001). AngII-induced vasoconstriction involves depolarization and VGCCs in afferent arteriole but ROCs and SOCs in efferent arteriole (Loutzenhiser and Loutzenhiser 2000). The vasoconstriction mechanisms in the renal medulla are similar to the preglomerular vasculature and involve depolarization and VGCCs (Hansen et al. 2001).

However, simultaneous measurements of force and [Ca^{2+}]$_i$ in vascular preparations loaded with the Ca^{2+} indicator aequorin or fura-2 have shown that the maintained agonist-induced contraction is associated with an initial [Ca^{2+}]$_i$ spike followed by a smaller increase in [Ca^{2+}]$_i$ (Morgan and Morgan 1984; Himpens et al. 1988; Khalil and van Breemen 1990). In contrast, membrane depolarization by high KCl solution causes maintained increases in both contraction and [Ca^{2+}]$_i$. Also, the agonist-induced [Ca^{2+}]$_i$-force relation is located to the left of that induced by high KCl, suggesting agonist-induced dissociations between [Ca^{2+}]$_i$ and force and increased myofilament force sensitivity to [Ca^{2+}]$_i$.

Agonist-induced dissociations between [Ca^{2+}]$_i$ and MLC phosphorylation have also been reported and have been explained by agonist-induced G protein-mediated change in the MLC kinase/MLC phosphatase activity ratio (Kitazawa et al. 2000; Somlyo and Somlyo 2000). Also, agonist-induced dissociations between MLC phosphorylation and force have been reported (Rembold 1990; Suematsu et al. 1991) and have been partly explained by the "latch bridge" hypothesis, which predicts that the dephosphorylation of myosin may generate a slowly cycling cross-bridge that supports force maintenance (Hai and Murphy 1989). However, in rabbit and ferret aorta, agonists cause a small and maintained contraction even in Ca^{2+}-free solution (Khalil and van Breemen 1988; Khalil and Morgan 1992) and in the absence of increases in $[Ca^{2+}]_i$ or MLC phosphorylation, suggesting activation of additional mechanisms that increase the myofilament force sensitivity to Ca^{2+} including protein kinase C and Rho kinase (Fig. 1.3).

Impaired VSM Ca^{2+} homeostasis has been suggested as a major cause of the increased vascular reactivity associated with HTN. Measurements of VSM $[Ca^{2+}]_i$ have shown that both basal and maintained agonist-induced $[Ca^{2+}]_i$ are greater in SHR than WKY (Sugiyama et al. 1986; Murphy and Khalil 2000). Agonist-stimulated $^{45}Ca^{2+}$ efflux and initial $[Ca^{2+}]_i$ transient in Ca^{2+}-free solution are not different in SHR and WKY, suggesting that the increased VSM $[Ca^{2+}]_i$ in HTN is not due to increased Ca^{2+} release from the sarcoplasmic reticulum (Cauvin and van Breemen 1985; Murphy and Khalil 2000). In contrast, the agonist-induced $^{45}Ca^{2+}$ influx and maintained $[Ca^{2+}]_i$ are greater in SHR than WKY, suggesting an increase in the permeability of plasmalemmal Ca^{2+} channels in SHR (Crews et al. 1999; Murphy and Khalil 2000). Also, Ca^{2+} extrusion mechanisms via the plasmalemmal Ca^{2+} pump may be compromised in SHR leading to accumulation of Ca^{2+} and increased $[Ca^{2+}]_i$ (Turla and Webb 1987).

Studies have examined possible vascular protection against HTN in females and whether these protective effects reflect differences in Ca^{2+} mobilization mechanisms of vasoconstriction (Crews et al. 1999). In rat aortic strips both phenylephrine and membrane depolarization by high KCl caused marked contraction and Ca^{2+} influx that were reduced in female compared with male WKY, and were greater in SHR than WKY in all groups of rats. The reduction in contraction and Ca^{2+} influx in females compared with males was greater in SHR than WKY. Also, the aortic contraction and Ca^{2+} influx were not different in ovariectomized female compared with intact male WKY or SHR or between ovariectomized female WKY or SHR with 17β-estradiol implant and intact females. These findings suggested that vascular reactivity and Ca^{2+} entry are dependent on gender and functional female gonads and that the gender-specific changes in vascular reactivity and Ca^{2+} influx are augmented in HTN (Crews et al. 1999).

Renal vascular reactivity to arginine-vasopressin (AVP) and AngII is increased in SHR (Hansen et al. 2005). The increased AVP-induced vascular reactivity and RVR are attributed to increased density of V1 receptor in renal interlobular arteries, and augmented VSM $[Ca^{2+}]_i$ due to Ca^{2+} mobilization from the intracellular stores and Ca^{2+} influx through plasma membrane (Fallet et al. 2005). In SHR, the enhanced vascular reactivity to AngII is due to increased amount/affinity of AT1R and may also involve increased basal $[Ca^{2+}]_i$, PKC activity and sensitivity of MLC to Ca^{2+}, and reduced cAMP production due to AngII-induced increase in formation of vasoconstrictive prostaglandins from arachidonic acid (Schiffrin 1994; Vyas and Jackson 1995; Ruan and Arendshorst 1996). Ca^{2+} signaling and store-operated Ca^{2+} entry are exaggerated in preglomerular VSM of young SHR (Fellner and Arendshorst 2002). Also, Na^+/Ca^{2+} exchanger and Ca^{2+} sequestering/extrusion mechanisms via sarcoplasmic reticulum and plasmalemmal Ca^{2+}-ATPase are diminished and may contribute to increased [Ca^{2+}]$_i$ in afferent arteriolar VSM of SHR (Nelson et al. 1996; Bell et al. 2000).

1.3.2.2 Protein Kinase C (PKC)

Diacylglycerol (DAG) is an important metabolic product of agonist-stimulated hydrolysis of PIP$_2$ and phosphatidylcholine (Fig. 1.3). DAG activates PKC, a family of at least 12 isoforms including conventional or classical PKC (cPKC) α, βI, βII, and γ, which require Ca^{2+}, phospholipid, and DAG; novel PKC (nPKC) δ, ε, η(L), θ, μ, and nu, which are Ca^{2+}-independent but require phospholipid and DAG; and atypical PKC (aPKC) ζ and λ/ι, which require phosphatidylserine but not DAG or Ca^{2+} for activation (Nishizuka 1992; Kanashiro and Khalil 1998b).

In several VSM preparations, phorbol esters bind to and activate PKC and induce maintained contraction (Danthuluri and Deth 1984; Khalil and van Breemen 1988) with no detectable increases in [Ca^{2+}]$_i$ or MLC phosphorylation (Jiang and Morgan 1987). Studies in ferret aorta VSMCs have suggested that this PKC-mediated contraction probably involves the Ca^{2+}-independent ε-PKC which increases the myofilament sensitivity to [Ca^{2+}]$_i$ (Collins et al. 1992; Horowitz et al. 1996).

A growing body of evidence suggests that agonists activate PKC during maintained VSM contraction. Agonists cause sustained increase in DAG, a known activator of PKC (Griendling et al. 1986). Also, PKC inhibitors such as staurosporine, chelerythrine, 1-(5-isoquinolinesulfonyl)-2-methylpiperazine (H-7), calphostin C, Gö6976, and the pseudosubstrate inhibitor peptide 19–36 inhibit agonist-induced VSM contraction (Khalil and van Breemen 1988; Collins et al. 1992). The role of Ca^{2+} release, Ca^{2+} influx, and PKC varies between agonists, with AngII depending on the three factors, TXA$_2$ independent of Ca^{2+}, and AVP having intermediate dependence (Cavarape et al. 2003). PKC plays a role in the regulation of renal vascular function and tubular transport. In rabbit, both afferent and efferent arterioles express Na$^+$/Ca^{2+} exchanger which is activated by PKC (Mitsuka and Berk 1991; Fowler et al. 1996).

PKC isoforms have different enzyme properties, substrates, functions, and subcellular distribution (Khalil et al. 1992; Liou and Morgan 1994). Ca^{2+}-dependent α- and β-PKC have been identified in many VSM preparations. Ca^{2+}-independent δ-, ε-, and ζ-PKC have been identified in VSM of rat aorta and mesenteric artery (Haller et al. 1994; Ohanian et al. 1996), ferret aorta (Andrea and Walsh 1992; Khalil et al. 1992), and porcine coronary artery (Kanashiro and Khalil 1998a). Ca^{2+}-dependent α-PKC is more abundant in ferret portal vein, while Ca^{2+}-independent ε-PKC is more abundant in ferret aorta (Andrea and Walsh 1992; Khalil et al. 1992; Khalil et al. 1994).

In resting cells, PKC is localized mainly in the cytosol. Activated PKC undergoes redistribution from the cytosolic to the particulate fraction. PKC α, β, and γ are mainly localized in the cytosolic fraction of resting cells and undergo translocation to the cell membranes in activated cells. δ-PKC is located almost exclusively in the particulate fraction and is associated with cytoskeletal proteins and may play a role in vascular remodeling. ζ-PKC is localized in the vicinity of the nucleus and may promote VSM growth and vascular hypertrophic changes associated with HTN (Khalil et al. 1992; Kanashiro and Khalil 1998b; Salamanca and Khalil 2005).

PKC phosphorylates a large number of substrates including CPI-17 protein, an inhibitor of MLC phosphatase, leading to increased phosphorylated/unphosphorylated MLC ratio (Kitazawa et al. 2000). PKC also phosphorylates the actin-binding protein calponin, leading to reversal of its inhibitory effects on myosin Mg^{2+}-ATPase and increased VSM contraction (Fig. 1.3).

The plasmalemmal location of certain PKC isoforms during agonist-induced activation of VSMCs has been difficult to reconcile with the central location of the contractile proteins, and possible involvement of a protein kinase cascade in PKC-mediated actions has been suggested. In resting ferret aorta VSMCs, mitogen-activated protein kinase (MAPK) is primarily distributed in the cytosol (Khalil and Morgan 1993). Cell activation is associated with a transient translocation of MAPK to the surface membrane followed by its redistribution back to the center of the cell during steady state. The redistribution of MAPK appears to be mediated by PKC as it follows PKC translocation and is prevented by the PKC-inhibitors staurosporine and calphostin C. PKC and MAPK may come together at the surface membrane as part of a complex of kinases and that PKC activation indirectly leads to activation of MAPK likely by MAPK kinase (MEK) (Childs et al. 1992) (Fig. 1.3). Caldesmon is an actin-binding protein (Ngai and Walsh 1984), and its phosphorylation by PKC-mediated activation of MAPK reverses its inhibitory effects on myosin Mg^{2+}-ATPase. MAPK is expressed in differentiated VSMCs and phosphorylates caldesmon, consistent with a role for MAPK in the signal transduction cascade linking PKC activation to VSM contraction (Adam et al. 1992; Childs et al. 1992).

PKC may play a role in the pathogenesis of HTN (Turla and Webb 1987; Liou and Morgan 1994). The enhanced vascular reactivity in SHR is associated with increased vascular PKC activity (Turla and Webb 1987; Bazan et al. 1992). Also, norepinephrine-induced contraction is more readily inhibited by the PKC-inhibitor H-7 in the aortas of SHR than WKY. Also, treatment of the aortic segments with H-7 caused a shift to the right in the concentration-contraction curve of the PKC activator TPA in the aortas of SHR, but not WKY (Shibata et al. 1990). The PKC activator PDBu produces increased contraction and greater reduction in cytosolic PKC activity in aortas from SHR than WKY, suggesting greater functional alterations of PKC in VSM of SHR (Bazan et al. 1992). In SHR, γ-interferon restores PKC level to the normal control rat, suggesting an interaction between PKC and the cytokine in HTN (Sauro and Hadden 1992).

To further evaluate the role of PKC in genetic HTN, studies have examined vascular contraction and PKC activity during the development of HTN in young (5–6 weeks) SHR. Contractions to high K^+ depolarizing solution in intact mesenteric arteries and the Ca^{2+}-force relationship in vessels permeabilized with α-toxin are not different in SHR and WKY rats. Treatment with the PKC activator PDBu augmented high K^+-induced contraction in intact vascular segments and enhanced the Ca^{2+}-force relationship in permeabilized vessels of SHR than WKY. Also, the PKC-inhibitors H-7 and calphostin C caused greater suppression of contraction in vascular segments of SHR than WKY. These data suggest that PKC enhances the

Ca^{2+} sensitivity of the contractile proteins in VSM and that the effects of PKC are greater in blood vessels of young prehypertensive SHR than WKY. The data also suggest that activation of PKC in VSM occurs before overt HTN and provide evidence for a role of PKC as a causative factor in the development of genetic HTN (Sasajima et al. 1997).

To further examine potential inborn differences in vascular PKC before the onset of HTN, studies have compared the proliferation of VSMCs from young (1–2 weeks) SHR and WKY rats. In cultured aortic VSM from SHR and WKY rats, both AngII and ET-1 enhanced thymidine incorporation into DNA, suggesting increased DNA synthesis. VSMC treatment with the PKC-inhibitor chelerythrine caused greater suppression of AngII and ET-1 induced DNA synthesis and VSMC growth in SHR than WKY, suggesting an inborn increase in PKC activity in VSMCs of SHR (Rosen et al. 1999).

Studies have also assessed the role of PKC in the changes in vascular tone associated with genetic HTN *in vivo* and examined the vascular effects of perfusing the PKC activator PDBu in the hind limb of anesthetized SHR and WKY rats. PDBu infusion into the hind limb caused prolonged vasoconstriction and elevation of perfusion pressure that were inhibited by the PKC-inhibitor staurosporine to a greater extent in SHR than WKY rats. These data provided evidence for a role of PKC in the regulation of vascular function and BP *in vivo* and further suggested an increase in PKC expression/activity in VSM of rat models of genetic HTN (Bilder et al. 1990).

α-PKC may play a role in Ca^{2+}-dependent contraction of VSM (Khalil et al. 1994) and overexpression of α-PKC has been implicated in HTN (Liou and Morgan 1994). In VSMCs of normotensive rats, α-PKC is localized mainly in the cytosol, but appears to be hyperactivated and concentrated at the cell membrane in VSMCs of hypertensive rats (Liou and Morgan 1994). Studies in cannulated rat femoral arterial branches have shown that high pressure induces $\cdot O_2^-$ production via PKC-dependent activation of NADPH oxidase (Ungvari et al. 2003). Also, the increased NADPH oxidase-mediated $\cdot O_2^-$ production in renovascular HTN involves PKC (Heitzer et al. 1999). MAPK activity in response to AngII and phorbol ester is enhanced in VSM from SHR, and inhibition of MAPK kinase attenuates AngII-mediated contraction in SHR VSM (Lucchesi et al. 1996; Touyz et al. 1999). Thus, increased PKC expression/activity in VSM may promote vasoconstriction and trophic vascular changes and lead to increased vascular resistance and HTN.

Gender differences in the expression/activity of PKC have been observed in aortic VSM of WKY and SHR. VSM contraction and the expression/activity of α-, δ-, and ζ-PKC in response to the phorbol ester PDBu are reduced in female compared with male WKY, and the gender-related differences are greater in VSM from SHR than WKY (Kanashiro and Khalil 2001). The PDBu-induced contraction and PKC activity were greater in ovariectomized than intact female rats. Treatment of OVX females with 17β-estradiol subcutaneous implants caused reduction in PDBu contraction and PKC activity that was more prominent in SHR than WKY. These data suggested gender-related reduction in VSM contraction and the expression/activity of α-, δ-, and ζ-PKC in females compared with males and that these differences are possibly mediated by estrogen and are enhanced in genetic forms of HTN (Kanashiro and Khalil 2001).

1.3.2.3 Rho Kinase

Some agonists such as TXA_2 activate the small G protein Rho and the Rho-kinase pathway. Rho kinase may contribute to VSM contraction by increasing the $[Ca^{2+}]_i$ sensitivity of contractile proteins (Hilgers and Webb 2005). Rho kinase inhibits MLC phosphatase leading to increased MLC phosphorylation and VSM contraction (Fig. 1.3). Arachidonic acid increases Rho-kinase activity, and Rho kinase may regulate the interaction of calponin with F-actin (Kaneko et al. 2000).

RhoA/Rho-kinase signaling plays a role in maintaining VSM contraction in response to stretch and receptor-mediated activation. NO causes vasodilation partly through inhibition of Rho kinase, and maintenance of renal basal vascular tone depends on a balance between NO production and Rho/Rho-kinase signaling. This is supported by the observation that RhoA/Rho-kinase signaling is amplified in eNOS knockout mouse (Sauzeau et al. 2003). Rho kinase is involved in the renal basal vascular tone, myogenic contraction of preglomerular and afferent arterioles, and vasoconstriction to AngII, TXA_2, and AVP (Nakamura et al. 2003). Also, the effect of reactive oxygen species (ROS) on vascular tone and VSM contraction may be mediated by Rho/Rho kinase. Increased NADPH-dependent ROS and reduced NO bioavailability activate Rho kinase which in turn inhibits MLC phosphatase, leading to increased MLC phosphorylation and VSM contraction (Jin et al. 2006).

Rho kinase-mediated Ca^{2+} sensitization may be involved in HTN. The Rho-kinase-inhibitor Y-27632 suppresses Rho-kinase-mediated formation of stress fibers in cultured cells and corrects HTN in hypertensive rat models (Uehata et al. 1997). While the expression of Rho GEF, a positive regulator of Rho kinase, is similar in WKY and young 4-week-old SHR, it is increased in SHR with established HTN, suggesting that increased Rho/Rho-kinase signaling may occur secondary to HTN (Ying et al. 2004). On the other hand, RhoA/Rho kinase may participate in the pathogenesis of nephrosclerosis in SHR, partly by upregulation of gene expressions of oxidative stress, ECM, adhesion molecules, and antifibrinolysis (Nishikimi et al. 2007).

1.3.2.4 Mitogen-Activated Protein Kinase (MAPK)

MAPK is one of several kinases that play a role in the transduction of extracellular mitogenic signals to the nucleus. MAPK is a Ser/Thr protein kinase that is fully activated by dual phosphorylation at Thr and Tyr residues and in turn activates nuclear transcription factors and VSMC growth and proliferation. In growth-arrested human VSMCs, MAPK is mainly cytosolic but translocates into the nucleus during activation by mitogens (Mii et al. 1996). In cultured VSMCs, AngII-induced activation of MAPK and mitogenic effects requires Ca^{2+}-dependent transactivation of epidermal growth factor receptor (EGFR) which provides docking sites for the upstream tyrosine kinase c-Src and the downstream adaptor proteins Shc and Grb2 that are essential for MAPK activation (Eguchi et al. 1998). MAPK in turn phosphorylates and activates several kinases and transcription factors and induces the nuclear proto-oncogene c-*fos* (Cobb and Goldsmith 1995). Tyrosine phosphorylation events involving EGFR, tyrosine kinase, and rapid c-Src activation also contribute to the renal microvascular and afferent arteriolar vasoconstriction and increased $[Ca^{2+}]_i$ and Ca^{2+} influx in response to AngII (Che and Carmines 2005). ET-1 induced activation of PKC, and MAPK also plays a role in VSM contraction and growth (Cain et al. 2002). In differentiated VSMCs, MAPK causes phosphorylation of the actin-binding protein caldesmon, thus allowing more actin to interact with myosin and thereby enhances VSM contraction (Salamanca and Khalil 2005).

MAPK-mediated VSM growth may underlie the vascular hypertrophic changes in HTN. MAPK activity is enhanced in the vasculature of SHR (Kubo et al. 1999). Also, the renal VSM expression/activity of EGFR is increased in Dahl salt-sensitive rats before the onset and in established HTN. EGFR mediates the fibrogenic and vasoconstrictor actions of AngII and ET-1, and inhibition of EGFR attenuates the HTN caused by AngII and the renal vascular fibrosis in L-NAME-induced HTN (Francois et al. 2004; Ying et al. 2004).

1.3.3 Extracellular Matrix and Vascular Remodeling in HTN

The vasculature undergoes significant remodeling, fibrotic changes, and structural alterations in HTN (Skov and Mulvany 2004). Renal vascular remodeling is associated with VSMC growth and proliferation and increased deposition of ECM proteins especially collagen type I, III, and IV in resistance arteries, glomerulus, and interstitium. NO inhibits collagen I gene activation, and chronic NOS inhibition is associated with activation of collagen I gene and accumulation of ECM in the renal vasculature. Also, AngII causes collagen I gene activation and fibrogenic action in the renal vasculature independent of hemodynamics and is likely mediated by ET-1, MAPK, and TGF-β (Tharaux et al. 2000). The fibrogenic action of ET-1 involves EGFR-mediated activation of MAPK p42/44 and collagen I gene (Francois et al. 2004). AngII-induced ET-1 expression in adventitial fibroblasts is partly mediated by NADPH oxidase, and ET-1 in turn stimulates collagen formation, implicating oxidative stress in vascular remodeling (An et al. 2007).

During vascular remodeling, matrix metalloproteinases (MMPs) degrade ECM proteins and adhesion molecules, thus enabling VSMCs to migrate and proliferate, and inflammatory cells to infiltrate the vessel wall (Visse and Nagase 2003) (Fig. 1.1). Endogenous tissue inhibitors of MMPs (TIMPs) provide a balancing mechanism and prevent excessive ECM degradation (Liu et al. 1997). Several MMPs are secreted in the vasculature in response to inflammatory cytokines and transcription factors and contribute to vascular remodeling by promoting ECM degradation. Additional effects of MMPs on ECs and VSM have been described (Raffetto and Khalil 2008).

HTN is associated with vascular remodeling and rearrangement of the vascular wall components and ECM proteins. Studies have examined the role of MMPs and TIMPs in the vascular remodeling associated with HTN. A clinical study in 44 hypertensive patients and 44 controls demonstrated that the plasma levels and activities of MMP-2, MMP-9, and TIMP-1 are increased in hypertensive patients, suggesting abnormal ECM metabolism (Derosa et al. 2006). Other studies have shown that the plasma levels of active MMP-2 and MMP-9 are depressed in patients with essential HTN. Also, a 6-month treatment with amlodipine normalized the plasma levels of MMP-9 but not MMP-2 (Zervoudaki et al. 2003). These studies supported a role of abnormal ECM metabolism in HTN and suggested that antihypertensive treatment may modulate collagen metabolism and that MMPs from the same family may have different effects on vascular function and vascular disease.

One study examined the serum levels of carboxy-terminal telopeptide of collagen type I (CITP) as a marker of extracellular collagen type I degradation, total matrix MMP-1 (collagenase), total TIMP-1, and MMP-1/TIMP-1 complex in 37 patients with never-treated HTN and in 23 normotensive control subjects. Serum levels of free MMP-1 and free TIMP-1 were calculated by subtracting the levels of MMP-1/TIMP-1 complex from the levels of total MMP-1 and total TIMP-1, respectively. Measurements were repeated in 26 hypertensive patients after 1 year of treatment with the ACE-inhibitor lisinopril. Interestingly, baseline free MMP-1 was decreased, and baseline free TIMP-1 was greater in hypertensive than normotensive subjects. No differences were observed in the baseline values of CITP between the two groups. Hypertensive patients with baseline left ventricular hypertrophy exhibited lower free MMP-1 and CITP and higher free TIMP-1 than hypertensive patients without baseline left ventricular hypertrophy. Treated patients showed an increase in free MMP-1 and a decrease in free TIMP-1. In addition, serum levels of CITP were greater in treated hypertensive patients than normotensive subjects. It was concluded that systemic extracellular degradation of collagen type I is depressed in HTN and can be normalized by treatment with lisinopril. A depressed degradation of collagen type I may facilitate organ fibrosis in hypertensive patients and those with left ventricular hypertrophy (Laviades et al. 1998).

Studies have also examined the expression/activity of MMPs in internal mammary artery specimens from normotensive and hypertensive patients undergoing coronary artery bypass surgery. Zymographic analysis indicated a decrease in total gelatinolytic activity of MMP-2 and MMP-9 in HTN. MMP-1 activity was also decreased by fourfold with no significant change in protein levels. Tissue levels of ECM inducer protein (EMMPRIN, a known stimulator of MMPs transcription), MMP activator protein (MT1-MMP), MMP-1, MMP-2, and MMP-9, as well as TIMP-1 and TIMP-2, were assessed by immunoblotting and revealed a decrease in MMP-9, EMMPRIN, and MT1-MMP levels in HTN. In addition, measurement of plasma markers of collagen synthesis (procollagen type I amino-terminal propeptide [PINP]) and collagen degradation (CITP) indicated no difference in PINP levels but suppressed degradation of collagen in HTN. These data demonstrate that not only MMP-1 and MMP-9 but MMP inducer and activator proteins are downregulated in the hypertensive state, resulting in enhanced collagen deposition in HTN (Ergul et al. 2004).

Experimental studies have also examined vascular remodeling and changes in the vascular tissue expression/activity of MMPs in animal models of HTN. Total wall and media thickness and MMP-2 expression and activity and TIMP-2 expression are increased in the aorta but not vena cava of DOCA-salt hypertensive vs. control rats. These data suggest a link between vascular remodeling in the aorta of DOCA-salt hypertensive rats and the actions of specific MMPs such as MMP-2 and that the increased TIMP-2 expression may be an adaptive increase to the higher-than-normal levels of MMP-2 (Watts et al. 2007).

Other studies have shown that in wild-type mice, AngII infusion 1 µg/kg/min by minipump plus a 5 % NaCl diet for 10 days is associated with HTN and increased MMP-9 activity in carotid artery. Interestingly, the absence of MMP-9 in MMP-$9^{(-/-)}$ mice was associated with vessel stiffness and increased pulse pressure, suggesting a beneficial role of MMP-9 activation in early HTN by preserving vessel compliance and alleviating the increase in BP (Flamant et al. 2007).

In SHR, all renal arteries except preglomerular arterioles exhibit increased media thickness and VSMC number and volume. Prehypertensive SHR also exhibits increased cross-sectional arterial media (Smeda et al. 1988b). In the vasculature of adult SHR, a decrease in MMP-1, MMP-2, and MMP-3 may contribute to remodeling of resistance arteries and the setting of HTN (Intengan and Schiffrin 2000).

1.4 Role of the Kidney

The kidney plays an important role in long-term control of BP through changes in extracellular fluid volume. A decrease in salt and water excretion by the kidney leads to salt and water retention, increased extracellular fluid volume, and increased blood volume. The increased blood volume would increase the venous return to the heart leading to increased preload, stroke volume CO, and BP (Fig. 1.1). The increase in BP and increased renal perfusion pressure allow the kidney to excrete the excess fluid volume. The renal-fluid volume mechanism has infinite feedback gain for controlling BP. The renal-body fluid-feedback mechanism for long-term control of BP predicts that persistent elevation of BP may occur due to either a reduction in renal sodium excretion or a hypertensive shift in the relationship between BP and sodium excretion (the pressure-natriuresis relationship) (Granger and Alexander 2000).

The kidney also controls BP and body fluid volume through RAS (Fig. 1.1). A decrease in BP stimulates renin release from the kidney. Renin cleaves angiotensinogen into AngI, which is further cleaved by a converting enzyme in the lung to AngII. AngII induces several cardiovascular effects including direct vasoconstriction (Touyz and Schiffrin 1997). AngII also stimulates VSMC growth and proliferation, leading to increased vascular resistance and BP (El Mabrouk et al. 2001; Yaghini et al. 2010). Additionally, circulating AngII could control body fluid volume by stimulating thirst and aldosterone secretion. Also, locally produced AngII has direct intrarenal actions that regulate renal hemodynamics and tubular reabsorption (Hall 1986; Hall et al. 1986) and could contribute to long-term BP regulation by promoting salt and water retention and by influencing pressure-natriuresis (Granger and Schnackenberg 2000).

The kidney plays an important role in the pathogenesis of HTN (Guyton 1987). BP returns to normal levels in hypertensive patients who receive kidneys from normotensive donors (Curtis et al. 1983). In hypertensive patients and animal models of HTN, the pressure-natriuresis relationship is shifted to the right and to a much higher pressure than for normotensive groups. In the non-salt-sensitive HTN, BP does not increase significantly with increases in salt intake. In the salt-sensitive HTN there is an additional change in the slope of the pressure-natriuresis relationship with increases in salt intake associated with marked increases in BP. Salt sensitivity may be related to abnormalities in the renal tissues and vessels. Inhibition of NO synthesis, as often seen in HTN, results in a hypertensive shift in renal pressure-natriuresis. The decrease in NO synthesis reduces renal sodium excretory function by acting directly on the renal vasculature and by altering tubular sodium transport.

Although vascular dysfunction, increased RVR, and changes in kidney function are demonstrated in several forms of HTN, whether their relationship is causative or associative in nature is unclear. The preglomerular vessels are resistance vessels and therefore may have the same alterations that occur in other peripheral resistance vessels in HTN. While this may be the case in some essential hypertensive patients, assessment of the renal hemodynamics in the prehypertensive state in humans suggest that an increase in RVR may constitute an initial event that leads to elevation of total vascular resistance and BP. Studies have shown an increase in RVR but normal renal blood flow (RBF) and GFR in young people with a family history of HTN as compared to control subjects, suggesting that abnormal renal vasoconstriction occurs in the prehypertensive state (Ruilope et al. 1994). Also, offsprings of hypertensive parents show an enhanced renal vasodilation in response to Ca^{2+} channel blockers, supporting the presence of functional renal vasoconstriction in subjects at risk of developing HTN (Ruilope et al. 1994). Borderline HTN is a condition in which a subject's BP is above normal but not sufficiently high to require antihypertensive therapy and may represent an early phase of essential HTN. Although RBF tends to be normal or slightly reduced in borderline HTN patients, RVR is elevated. Borderline hypertensives also exhibit an enhanced renal vasoconstrictor response to infusion of norepinephrine. Thus, while increased preglomerular vascular tone may be due to a generalized disturbance in peripheral vascular resistance, it could also be due to localized alterations in the renal vasculature and could lead to HTN (Ruilope et al. 1994).

Experimental studies also support a primary importance of renal vascular alterations in the development of HTN. Experimental cross transplantation studies suggest that the underlying pathology of HTN travels with the kidney from SHR donors to normotensive recipients, and kidneys from normotensive donors lower BP in transplanted SHR (Curtis et al. 1983; Grisk and Rettig 2001). Measurement of the renal hemodynamics in SHR showed increased RVR as HTN develops while GFR and glomerular capillary pressure are normal, and the unaltered glomerular function may be in part due to a renal autoregulatory adjustment to the increase in BP. However, in young SHR, an increase in RVR and renal vascular hypertrophy occur as early as 4 weeks of age during which there is no established HTN (Smeda et al. 1988b). Also, in prehypertensive young SHR, there is increased reactivity of the renal vasculature to vasoconstrictors and decreased medullary blood flow, supporting a role in the development of HTN (Vyas and Jackson 1995; Sasajima et al. 1997). Furthermore, in SHR, preglomerular arterial wall hypertrophy is not reversed by antihypertensive treatment, which is not the case in other vascular beds (Smeda et al. 1988a). The presence of structural and functional changes in preglomerular and afferent arterioles before the development of HTN and the persistence of structural changes despite normalization of BP suggest that these renal vascular changes are not a mere consequence of elevated BP but may be involved in the pathogenesis of HTN.

1.5 Perspective

The present review casts highlights on the roles of the heart, blood vessels, and the kidney in the control of BP and how dysfunction in any of these systems could cause HTN. However, the diverse hemodynamic and pathophysiologic changes associated with HTN are more likely the result of more than a single cause. In many cases vascular dysfunction could lead to changes in cardiac and renal function or vice versa. Combined changes in cardiac, vascular, and renal function are more common and are often associated with persistent increases in BP and HTN. Also, whether the changes in EC and VSM and the ensuing changes in vascular reactivity represent the cause or consequence of HTN remains unclear. Because of the multifactorial nature of HTN, several lines of treatment may be needed. Inhibitors of RAS have been effective in treatment of HTN. However, ET-1 antagonists may not be beneficial in all forms of HTN. Ca^{2+} channel antagonists may be the drugs of choice to control the increased vascular reactivity caused by increased Ca^{2+} influx. However, Ca^{2+} antagonist-insensitive forms of VSM contraction have been identified and suggested the contribution of other signaling pathways in addition to increased $[Ca^{2+}]_i$ in the pathogenesis of HTN. Selective PKC inhibitors may reverse PKC-dependent increases in vascular reactivity. Inhibition of MAPK or Rho kinase represents other new and exciting areas of research. Some of the newly developed MAPK kinase inhibitors, such as PD 98059, and Rho kinase inhibitors, such as Y-27632, appear to be selective, but may need further evaluation before they can be used safely in treating HTN in humans.

Acknowledgments This work was supported by grants from the National Heart, Lung, and Blood Institute (HL-65998, HL-98724) and The Eunice Kennedy Shriver National Institute of Child Health and Human Development (HD-60702).

References

Adam LP, Gapinski CJ, Hathaway DR (1992) Phosphorylation sequences in h-caldesmon from phorbol ester-stimulated canine aortas. FEBS Lett 302:223–226

Alonso-Galicia M, Maier KG, Greene AS, Cowley AW Jr, Roman RJ (2002) Role of 20-hydroxyeicosatetraenoic acid in the renal and vasoconstrictor actions of angiotensin II. Am J Physiol Regul Integr Comp Physiol 283:R60–R68

Amerena J, Julius S (1995) The role of the autonomic nervous system in hypertension. Hypertens Res 18:99–110

An SJ, Boyd R, Zhu M, Chapman A, Pimentel DR, Wang HD (2007) NADPH oxidase mediates angiotensin II-induced endothelin-1 expression in vascular adventitial fibroblasts. Cardiovasc Res 75:702–709

Andrea JE, Walsh MP (1992) Protein kinase C of smooth muscle. Hypertension 20:585–595

Anning PB, Coles B, Bermudez-Fajardo A, Martin PE, Levison BS, Hazen SL, Funk CD, Kuhn H, O'Donnell VB (2005) Elevated endothelial nitric oxide bioactivity and resistance to angiotensin-dependent hypertension in 12/15-lipoxygenase knockout mice. Am J Pathol 166: 653–662

Barman SA (2007) Vasoconstrictor effect of endothelin-1 on hypertensive pulmonary arterial smooth muscle involves Rho-kinase and protein kinase C. Am J Physiol Lung Cell Mol Physiol 293:L472–L479

Baylis C, Qiu C (1996) Importance of nitric oxide in the control of renal hemodynamics. Kidney Int 49:1727–1731

Bazan E, Campbell AK, Rapoport RM (1992) Protein kinase C activity in blood vessels from normotensive and spontaneously hypertensive rats. Eur J Pharmacol 227:343–348

Bell PD, Mashburn N, Unlap MT (2000) Renal sodium/calcium exchange; a vasodilator that is defective in salt-sensitive hypertension. Acta Physiol Scand 168:209–214

Bilder GE, Kasiewski CJ, Perrone MH (1990) Phorbol-12,13-dibutyrate-induced vasoconstriction in vivo: characterization of response in genetic hypertension. J Pharmacol Exp Ther 252:526–530

Boulanger CM (1999) Secondary endothelial dysfunction: hypertension and heart failure. J Mol Cell Cardiol 31:39–49

Brannstrom K, Arendshorst WJ (1999) Thromboxane A2 contributes to the enhanced tubuloglomerular feedback activity in young SHR. Am J Physiol 276:F758–F766

Brovkovych V, Stolarczyk E, Oman J, Tomboulian P, Malinski T (1999) Direct electrochemical measurement of nitric oxide in vascular endothelium. J Pharm Biomed Anal 19: 135–143

Busse R, Edwards G, Feletou M, Fleming I, Vanhoutte PM, Weston AH (2002) EDHF: bringing the concepts together. Trends Pharmacol Sci 23:374–380

Bussemaker E, Popp R, Fisslthaler B, Larson CM, Fleming I, Busse R, Brandes RP (2003) Aged spontaneously hypertensive rats exhibit a selective loss of EDHF-mediated relaxation in the renal artery. Hypertension 42:562–568

Cai H, Harrison DG (2000) Endothelial dysfunction in cardiovascular diseases: the role of oxidant stress. Circ Res 87:840–844

Cain AE, Tanner DM, Khalil RA (2002) Endothelin-1 – induced enhancement of coronary smooth muscle contraction via MAPK-dependent and MAPK-independent [Ca(2+)](i) sensitization pathways. Hypertension 39:543–549

Cardillo C, Panza JA (1998) Impaired endothelial regulation of vascular tone in patients with systemic arterial hypertension. Vasc Med 3:138–144

Cardillo C, Kilcoyne CM, Quyyumi AA, Cannon RO 3rd, Panza JA (1998) Selective defect in nitric oxide synthesis may explain the impaired endothelium-dependent vasodilation in patients with essential hypertension. Circulation 97:851–856

Cauvin C, van Breemen C (1985) Different Ca2+ channels along the arterial tree. J Cardiovasc Pharmacol 7(Suppl 4):S4–S10

Cavarape A, Bauer J, Bartoli E, Endlich K, Parekh N (2003) Effects of angiotensin II, arginine vasopressin and tromboxane A2 in renal vascular bed: role of rho-kinase. Nephrol Dial Transplant 18:1764–1769

Chapleau CE, White RP (1979) Effects of prostacyclin on the canine isolated basilar artery. Prostaglandins 17:573–580

Chatziantoniou C, Daniels FH, Arendshorst WJ (1990) Exaggerated renal vascular reactivity to angiotensin and thromboxane in young genetically hypertensive rats. Am J Physiol 259:F372–F382

Che Q, Carmines PK (2005) Src family kinase involvement in rat preglomerular microvascular contractile and [Ca2+]i responses to ANG II. Am J Physiol Renal Physiol 288:F658–F664

Childs TJ, Watson MH, Sanghera JS, Campbell DL, Pelech SL, Mak AS (1992) Phosphorylation of smooth muscle caldesmon by mitogen-activated protein (MAP) kinase and expression of MAP kinase in differentiated smooth muscle cells. J Biol Chem 267:22853–22859

Chou TC, Yen MH, Li CY, Ding YA (1998) Alterations of nitric oxide synthase expression with aging and hypertension in rats. Hypertension 31:643–648

Cobb MH, Goldsmith EJ (1995) How MAP kinases are regulated. J Biol Chem 270:14843–14846

Collins EM, Walsh MP, Morgan KG (1992) Contraction of single vascular smooth muscle cells by phenylephrine at constant [Ca2+]i. Am J Physiol 262:H754–H762

Crews JK, Murphy JG, Khalil RA (1999) Gender differences in Ca(2+) entry mechanisms of vasoconstriction in Wistar-Kyoto and spontaneously hypertensive rats. Hypertension 34:931–936

Curtis JJ, Luke RG, Dustan HP, Kashgarian M, Whelchel JD, Jones P, Diethelm AG (1983) Remission of essential hypertension after renal transplantation. N Engl J Med 309:1009–1015

Danthuluri NR, Deth RC (1984) Phorbol ester-induced contraction of arterial smooth muscle and inhibition of alpha-adrenergic response. Biochem Biophys Res Commun 125:1103–1109

De Artinano AA, Gonzalez VL (1999) Endothelial dysfunction and hypertensive vasoconstriction. Pharmacol Res 40:113–124

de Groot MC, Illing B, Horn M, Urban B, Haase A, Schnackerz K, Neubauer S (1998) Endothelin-1 increases susceptibility of isolated rat hearts to ischemia/reperfusion injury by reducing coronary flow. J Mol Cell Cardiol 30:2657–2668

De Vriese AS, Van de Voorde J, Lameire NH (2002) Effects of connexin-mimetic peptides on nitric oxide synthase- and cyclooxygenase-independent renal vasodilation. Kidney Int 61:177–185

de Wit C, Roos F, Bolz SS, Kirchhoff S, Kruger O, Willecke K, Pohl U (2000) Impaired conduction of vasodilation along arterioles in connexin40-deficient mice. Circ Res 86:649–655

Derosa G, D'Angelo A, Ciccarelli L, Piccinni MN, Pricolo F, Salvadeo S, Montagna L, Gravina A, Ferrari I, Galli S, Paniga S, Tinelli C, Cicero AF (2006) Matrix metalloproteinase-2, -9, and tissue inhibitor of metalloproteinase-1 in patients with hypertension. Endothelium 13:227–231

Dubinion JH, Mi Z, Jackson EK (2006) Role of renal sympathetic nerves in regulating renovascular responses to angiotensin II in spontaneously hypertensive rats. J Pharmacol Exp Ther 317:1330–1336

Eguchi S, Numaguchi K, Iwasaki H, Matsumoto T, Yamakawa T, Utsunomiya H, Motley ED, Kawakatsu H, Owada KM, Hirata Y, Marumo F, Inagami T (1998) Calcium-dependent epidermal growth factor receptor transactivation mediates the angiotensin II-induced mitogen-activated protein kinase activation in vascular smooth muscle cells. J Biol Chem 273:8890–8896

El Mabrouk M, Touyz RM, Schiffrin EL (2001) Differential ANG II-induced growth activation pathways in mesenteric artery smooth muscle cells from SHR. Am J Physiol Heart Circ Physiol 281:H30–H39

Ergul A, Portik-Dobos V, Hutchinson J, Franco J, Anstadt MP (2004) Downregulation of vascular matrix metalloproteinase inducer and activator proteins in hypertensive patients. Am J Hypertens 17:775–782

Fagard R (1992) Hypertensive heart disease: pathophysiology and clinical and prognostic consequences. J Cardiovasc Pharmacol 19(Suppl 5):S59–S66

Fallet RW, Ikenaga H, Bast JP, Carmines PK (2005) Relative contributions of Ca2+ mobilization and influx in renal arteriolar contractile responses to arginine vasopressin. Am J Physiol Renal Physiol 288:F545–F551

Falloon BJ, Heagerty AM (1994) In vitro perfusion studies of human resistance artery function in essential hypertension. Hypertension 24:16–23

Feletou M, Vanhoutte PM (2006) Endothelium-derived hyperpolarizing factor: where are we now? Arterioscler Thromb Vasc Biol 26:1215–1225

Feletou M, Vanhoutte PM (2009) EDHF: an update. Clin Sci (Lond) 117:139–155

Fellner SK, Arendshorst WJ (2002) Store-operated Ca2+ entry is exaggerated in fresh preglomerular vascular smooth muscle cells of SHR. Kidney Int 61:2132–2141

Flamant M, Placier S, Dubroca C, Esposito B, Lopes I, Chatziantoniou C, Tedgui A, Dussaule JC, Lehoux S (2007) Role of matrix metalloproteinases in early hypertensive vascular remodeling. Hypertension 50:212–218

Fleming I, Busse R (1999) NO: the primary EDRF. J Mol Cell Cardiol 31:5–14

Floras JS, Hara K (1993) Sympathoneural and haemodynamic characteristics of young subjects with mild essential hypertension. J Hypertens 11:647–655

Fouad-Tarazi FM (1988) Factors contributing to resistant hypertension. Cardiac considerations. Hypertension 11:II84–II87

Fowler BC, Carmines PK, Nelson LD, Bell PD (1996) Characterization of sodium-calcium exchange in rabbit renal arterioles. Kidney Int 50:1856–1862

Franco M, Tapia E, Santamaria J, Zafra I, Garcia-Torres R, Gordon KL, Pons H, Rodriguez-Iturbe B, Johnson RJ, Herrera-Acosta J (2001) Renal cortical vasoconstriction contributes to development of salt-sensitive hypertension after angiotensin II exposure. J Am Soc Nephrol 12:2263–2271

Franco M, Martinez F, Quiroz Y, Galicia O, Bautista R, Johnson RJ, Rodriguez-Iturbe B (2007) Renal angiotensin II concentration and interstitial infiltration of immune cells are correlated with blood pressure levels in salt-sensitive hypertension. Am J Physiol Regul Integr Comp Physiol 293:R251–R256

Francois H, Placier S, Flamant M, Tharaux PL, Chansel D, Dussaule JC, Chatziantoniou C (2004) Prevention of renal vascular and glomerular fibrosis by epidermal growth factor receptor inhibition. FASEB J 18:926–928

Frohlich ED, Kozul VJ, Tarazi RC, Dustan HP (1970) Physiological comparison of labile and essential hypertension. Circ Res 27:55–69

Furchgott RF, Vanhoutte PM (1989) Endothelium-derived relaxing and contracting factors. FASEB J 3:2007–2018

Gluais P, Lonchampt M, Morrow JD, Vanhoutte PM, Feletou M (2005) Acetylcholine-induced endothelium-dependent contractions in the SHR aorta: the Janus face of prostacyclin. Br J Pharmacol 146:834–845

Goddard J, Webb DJ (2000) Plasma endothelin concentrations in hypertension. J Cardiovasc Pharmacol 35:S25–S31

Granger JP, Alexander BT (2000) Abnormal pressure-natriuresis in hypertension: role of nitric oxide. Acta Physiol Scand 168:161–168

Granger JP, Schnackenberg CG (2000) Renal mechanisms of angiotensin II-induced hypertension. Semin Nephrol 20:417–425

Griendling KK, Rittenhouse SE, Brock TA, Ekstein LS, Gimbrone MA Jr, Alexander RW (1986) Sustained diacylglycerol formation from inositol phospholipids in angiotensin II-stimulated vascular smooth muscle cells. J Biol Chem 261:5901–5906

Grisk O, Rettig R (2001) Renal transplantation studies in genetic hypertension. News Physiol Sci 16:262–265

Gruetter CA, Gruetter DY, Lyon JE, Kadowitz PJ, Ignarro LJ (1981) Relationship between cyclic guanosine 3′:5′-monophosphate formation and relaxation of coronary arterial smooth muscle by glyceryl trinitrate, nitroprusside, nitrite and nitric oxide: effects of methylene blue and methemoglobin. J Pharmacol Exp Ther 219:181–186

Guyton AC (1987) Renal function curve – a key to understanding the pathogenesis of hypertension. Hypertension 10:1–6

Hai CM, Murphy RA (1989) Ca2+, crossbridge phosphorylation, and contraction. Annu Rev Physiol 51:285–298

Hall JE (1986) Control of sodium excretion by angiotensin II: intrarenal mechanisms and blood pressure regulation. Am J Physiol 250:R960–R972

Hall JE, Mizelle HL, Woods LL (1986) The renin-angiotensin system and long-term regulation of arterial pressure. J Hypertens 4:387–397

Haller H (1996) Hypertension, the endothelium and the pathogenesis of chronic vascular disease. Kidney Blood Press Res 19:166–171

Haller H, Quass P, Lindschau C, Luft FC, Distler A (1994) Platelet-derived growth factor and angiotensin II induce different spatial distribution of protein kinase C-alpha and -beta in vascular smooth muscle cells. Hypertension 23:848–852

Hansen PB, Jensen BL, Andreasen D, Skott O (2001) Differential expression of T- and L-type voltage-dependent calcium channels in renal resistance vessels. Circ Res 89:630–638

Hansen FH, Vagnes OB, Iversen BM (2005) Enhanced response to AVP in the interlobular artery from the spontaneously hypertensive rat. Am J Physiol Renal Physiol 288:F1023–F1031

Hedner T, Sun X (1997) Measures of endothelial function as an endpoint in hypertension? Blood Press Suppl 2:58–66

Heitzer T, Wenzel U, Hink U, Krollner D, Skatchkov M, Stahl RA, MacHarzina R, Brasen JH, Meinertz T, Munzel T (1999) Increased NAD(P)H oxidase-mediated superoxide production in renovascular hypertension: evidence for an involvement of protein kinase C. Kidney Int 55:252–260

Hilgers RH, Webb RC (2005) Molecular aspects of arterial smooth muscle contraction: focus on Rho. Exp Biol Med (Maywood) 230:829–835

Himpens B, Matthijs G, Somlyo AV, Butler TM, Somlyo AP (1988) Cytoplasmic free calcium, myosin light chain phosphorylation, and force in phasic and tonic smooth muscle. J Gen Physiol 92:713–729

Horowitz A, Menice CB, Laporte R, Morgan KG (1996) Mechanisms of smooth muscle contraction. Physiol Rev 76:967–1003

Huang PL, Huang Z, Mashimo H, Bloch KD, Moskowitz MA, Bevan JA, Fishman MC (1995) Hypertension in mice lacking the gene for endothelial nitric oxide synthase. Nature 377:239–242

Ignarro LJ, Kadowitz PJ (1985) The pharmacological and physiological role of cyclic GMP in vascular smooth muscle relaxation. Annu Rev Pharmacol Toxicol 25:171–191

Ignarro LJ, Buga GM, Wood KS, Byrns RE, Chaudhuri G (1987) Endothelium-derived relaxing factor produced and released from artery and vein is nitric oxide. Proc Natl Acad Sci USA 84:9265–9269

Imig JD (2005) Epoxide hydrolase and epoxygenase metabolites as therapeutic targets for renal diseases. Am J Physiol Renal Physiol 289:F496–F503

Intengan HD, Schiffrin EL (2000) Structure and mechanical properties of resistance arteries in hypertension: role of adhesion molecules and extracellular matrix determinants. Hypertension 36:312–318

Jameson M, Dai FX, Luscher T, Skopec J, Diederich A, Diederich D (1993) Endothelium-derived contracting factors in resistance arteries of young spontaneously hypertensive rats before development of overt hypertension. Hypertension 21:280–288

Jiang MJ, Morgan KG (1987) Intracellular calcium levels in phorbol ester-induced contractions of vascular muscle. Am J Physiol 253:H1365–H1371

Jin L, Ying Z, Hilgers RH, Yin J, Zhao X, Imig JD, Webb RC (2006) Increased RhoA/Rho-kinase signaling mediates spontaneous tone in aorta from angiotensin II-induced hypertensive rats. J Pharmacol Exp Ther 318:288–295

Johnson RJ, Alpers CE, Yoshimura A, Lombardi D, Pritzl P, Floege J, Schwartz SM (1992) Renal injury from angiotensin II-mediated hypertension. Hypertension 19:464–474

Johnson FK, Johnson RA, Peyton KJ, Durante W (2005) Arginase inhibition restores arteriolar endothelial function in Dahl rats with salt-induced hypertension. Am J Physiol Regul Integr Comp Physiol 288:R1057–R1062

Julius S, Pascual AV, Sannerstedt R, Mitchell C (1971) Relationship between cardiac output and peripheral resistance in borderline hypertension. Circulation 43:382–390

Jun T, Ke-yan F, Catalano M (1996) Increased superoxide anion production in humans: a possible mechanism for the pathogenesis of hypertension. J Hum Hypertens 10:305–309

Kamm KE, Stull JT (1989) Regulation of smooth muscle contractile elements by second messengers. Annu Rev Physiol 51:299–313

Kanashiro CA, Khalil RA (1998a) Isoform-specific protein kinase C activity at variable Ca2+ entry during coronary artery contraction by vasoactive eicosanoids. Can J Physiol Pharmacol 76:1110–1119

Kanashiro CA, Khalil RA (1998b) Signal transduction by protein kinase C in mammalian cells. Clin Exp Pharmacol Physiol 25:974–985

Kanashiro CA, Khalil RA (2001) Gender-related distinctions in protein kinase C activity in rat vascular smooth muscle. Am J Physiol Cell Physiol 280:C34–C45

Kaneko T, Amano M, Maeda A, Goto H, Takahashi K, Ito M, Kaibuchi K (2000) Identification of calponin as a novel substrate of Rho-kinase. Biochem Biophys Res Commun 273:110–116

Kawai Y, Ohhashi T (1994) Effects of isocarbacyclin, a stable prostacyclin analogue, on monkey isolated cerebral and peripheral arteries. Br J Pharmacol 112:635–639

Khalil RA, Morgan KG (1992) Phenylephrine-induced translocation of protein kinase C and shortening of two types of vascular cells of the ferret. J Physiol 455:585–599

Khalil RA, Morgan KG (1993) PKC-mediated redistribution of mitogen-activated protein kinase during smooth muscle cell activation. Am J Physiol 265:C406–C411

Khalil RA, van Breemen C (1988) Sustained contraction of vascular smooth muscle: calcium influx or C-kinase activation? J Pharmacol Exp Ther 244:537–542

Khalil RA, van Breemen C (1990) Intracellular free calcium concentration/force relationship in rabbit inferior vena cava activated by norepinephrine and high K+. Pflugers Arch 416:727–734

Khalil RA, Lajoie C, Resnick MS, Morgan KG (1992) Ca(2+)-independent isoforms of protein kinase C differentially translocate in smooth muscle. Am J Physiol 263:C714–C719

Khalil RA, Lajoie C, Morgan KG (1994) In situ determination of [Ca2+]i threshold for translocation of the alpha-protein kinase C isoform. Am J Physiol 266:C1544–C1551

Kim YS, Xu ZG, Reddy MA, Li SL, Lanting L, Sharma K, Adler SG, Natarajan R (2005) Novel interactions between TGF-{beta}1 actions and the 12/15-lipoxygenase pathway in mesangial cells. J Am Soc Nephrol 16:352–362

Kiowski W (1999) Endothelial dysfunction in hypertension. Clin Exp Hypertens 21:635–646

Kitazawa T, Eto M, Woodsome TP, Brautigan DL (2000) Agonists trigger G protein-mediated activation of the CPI-17 inhibitor phosphoprotein of myosin light chain phosphatase to enhance vascular smooth muscle contractility. J Biol Chem 275:9897–9900

Kobori H, Nishiyama A, Abe Y, Navar LG (2003) Enhancement of intrarenal angiotensinogen in Dahl salt-sensitive rats on high salt diet. Hypertension 41:592–597

Kubo T, Ibusuki T, Saito E, Kambe T, Hagiwara Y (1999) Vascular mitogen-activated protein kinase activity is enhanced via angiotensin system in spontaneously hypertensive rats. Eur J Pharmacol 372:279–285

Laviades C, Varo N, Fernandez J, Mayor G, Gil MJ, Monreal I, Diez J (1998) Abnormalities of the extracellular degradation of collagen type I in essential hypertension. Circulation 98:535–540

Li L, Bressler B, Prameya R, Dorovini-Zis K, Van Breemen C (1999) Agonist-stimulated calcium entry in primary cultures of human cerebral microvascular endothelial cells. Microvasc Res 57:211–226

Liou YM, Morgan KG (1994) Redistribution of protein kinase C isoforms in association with vascular hypertrophy of rat aorta. Am J Physiol 267:C980–C989

Liu FY, Cogan MG (1988) Angiotensin II stimulation of hydrogen ion secretion in the rat early proximal tubule. Modes of action, mechanism, and kinetics. J Clin Invest 82:601–607

Liu YE, Wang M, Greene J, Su J, Ullrich S, Li H, Sheng S, Alexander P, Sang QA, Shi YE (1997) Preparation and characterization of recombinant tissue inhibitor of metalloproteinase 4 (TIMP-4). J Biol Chem 272:20479–20483

Loutzenhiser K, Loutzenhiser R (2000) Angiotensin II-induced Ca(2+) influx in renal afferent and efferent arterioles: differing roles of voltage-gated and store-operated Ca(2+) entry. Circ Res 87:551–557

Lucchesi PA, Bell JM, Willis LS, Byron KL, Corson MA, Berk BC (1996) Ca(2+)-dependent mitogen-activated protein kinase activation in spontaneously hypertensive rat vascular smooth muscle defines a hypertensive signal transduction phenotype. Circ Res 78:962–970

Lund-Johansen P (1986) Hemodynamic patterns in the natural history of borderline hypertension. J Cardiovasc Pharmacol 8(Suppl 5):S8–S14

Lund-Johansen P (1987) Hemodynamics in hypertension at rest and during exercise. J Cardiovasc Pharmacol 10(Suppl 11):S1–S5

Lund-Johansen P (1991) Twenty-year follow-up of hemodynamics in essential hypertension during rest and exercise. Hypertension 18:III54–III61

Lund-Johansen P (1993) Relationship between cardiovascular haemodynamics and goals of antihypertensive therapy. J Hum Hypertens 7(Suppl 1):S21–S28

Lund-Johansen P, Omvik P (1990) The role of multiple action agents in hypertension. Eur J Clin Pharmacol 38(Suppl 2):S89–S95

Luscher TF (1990) Endothelial control of vascular tone and growth. Clin Exp Hypertens A 12:897–902

Luscher TF (1992) Heterogeneity of endothelial dysfunction in hypertension. Eur Heart J 13 Suppl D:50–55

Luscher TF (1994) The endothelium in hypertension: bystander, target or mediator? J Hypertens Suppl 12:S105–S116

Luscher TF, Noll G (1995) The pathogenesis of cardiovascular disease: role of the endothelium as a target and mediator. Atherosclerosis 118(Suppl):S81–S90

Luscher TF, Dohi Y, Tanner FC, Boulanger C (1991) Endothelium-dependent control of vascular tone: effects of age, hypertension and lipids. Basic Res Cardiol 86(Suppl 2):143–158

Luscher TF, Dohi Y, Tschudi M (1992a) Endothelium-dependent regulation of resistance arteries: alterations with aging and hypertension. J Cardiovasc Pharmacol 19(Suppl 5):S34–S42

Luscher TF, Tanner FC, Dohi Y (1992b) Age, hypertension and hypercholesterolaemia alter endothelium-dependent vascular regulation. Pharmacol Toxicol 70:S32–S39

Lutas EM, Devereux RB, Reis G, Alderman MH, Pickering TG, Borer JS, Laragh JH (1985) Increased cardiac performance in mild essential hypertension. Left ventricular mechanics. Hypertension 7:979–988

Ma YH, Gebremedhin D, Schwartzman ML, Falck JR, Clark JE, Masters BS, Harder DR, Roman RJ (1993) 20-Hydroxyeicosatetraenoic acid is an endogenous vasoconstrictor of canine renal arcuate arteries. Circ Res 72:126–136

Majed BH, Khalil RA (2012) Molecular mechanisms regulating the vascular prostacyclin pathways and their adaptation during pregnancy and in the newborn. Pharmacol Rev 64:540–582

Matsumura Y, Hashimoto N, Taira S, Kuro T, Kitano R, Ohkita M, Opgenorth TJ, Takaoka M (1999) Different contributions of endothelin-A and endothelin-B receptors in the pathogenesis of deoxycorticosterone acetate-salt-induced hypertension in rats. Hypertension 33:759–765

McAllister AS, Atkinson AB, Johnston GD, Hadden DR, Bell PM, McCance DR (1999) Basal nitric oxide production is impaired in offspring of patients with essential hypertension. Clin Sci (Lond) 97:141–147

McCulloch KM, Docherty C, MacLean MR (1998) Endothelin receptors mediating contraction of rat and human pulmonary resistance arteries: effect of chronic hypoxia in the rat. Br J Pharmacol 123:1621–1630

Messerli FH, Sundgaard-Riise K, Ventura HO, Dunn FG, Glade LB, Frohlich ED (1983) Essential hypertension in the elderly: haemodynamics, intravascular volume, plasma renin activity, and circulating catecholamine levels. Lancet 2:983–986

Mii S, Khalil RA, Morgan KG, Ware JA, Kent KC (1996) Mitogen-activated protein kinase and proliferation of human vascular smooth muscle cells. Am J Physiol 270:H142–H150

Miller SB (2006) Prostaglandins in health and disease: an overview. Semin Arthritis Rheum 36:37–49

Mitsuka M, Berk BC (1991) Long-term regulation of Na(+)-H+ exchange in vascular smooth muscle cells: role of protein kinase C. Am J Physiol 260:C562–C569

Modesti PA, Cecioni I, Costoli A, Poggesi L, Galanti G, Serneri GG (2000) Renal endothelin in heart failure and its relation to sodium excretion. Am Heart J 140:617–622

Morgan JP, Morgan KG (1984) Stimulus-specific patterns of intracellular calcium levels in smooth muscle of ferret portal vein. J Physiol 351:155–167

Mulvany MJ, Nyborg N (1980) An increased calcium sensitivity of mesenteric resistance vessels in young and adult spontaneously hypertensive rats. Br J Pharmacol 71:585–596

Murphy JG, Khalil RA (2000) Gender-specific reduction in contractility and [Ca(2+)](i) in vascular smooth muscle cells of female rat. Am J Physiol Cell Physiol 278:C834–C844

Murphy JG, Herrington JN, Granger JP, Khalil RA (2003) Enhanced [Ca2+]i in renal arterial smooth muscle cells of pregnant rats with reduced uterine perfusion pressure. Am J Physiol Heart Circ Physiol 284:H393–H403

Nakamura A, Hayashi K, Ozawa Y, Fujiwara K, Okubo K, Kanda T, Wakino S, Saruta T (2003) Vessel- and vasoconstrictor-dependent role of rho/rho-kinase in renal microvascular tone. J Vasc Res 40:244–251

Naruse M, Naruse K, Yoshimoto T, Tanaka M, Tanabe A, Demura H (1994) Clinical significance of nitric oxide in hypertension. Nihon Naibunpi Gakkai Zasshi 70:489–502

Nava E, Luscher TF (1995) Endothelium-derived vasoactive factors in hypertension: nitric oxide and endothelin. J Hypertens Suppl 13:S39–S48

Nelson LD, Mashburn NA, Bell PD (1996) Altered sodium-calcium exchange in afferent arterioles of the spontaneously hypertensive rat. Kidney Int 50:1889–1896

Ngai PK, Walsh MP (1984) Inhibition of smooth muscle actin-activated myosin Mg2+-ATPase activity by caldesmon. J Biol Chem 259:13656–13659

Nguyen PV, Parent A, Deng LY, Fluckiger JP, Thibault G, Schiffrin EL (1992) Endothelin vascular receptors and responses in deoxycorticosterone acetate-salt hypertensive rats. Hypertension 19:II98–II104

Nishikimi T, Koshikawa S, Ishikawa Y, Akimoto K, Inaba C, Ishimura K, Ono H, Matsuoka H (2007) Inhibition of Rho-kinase attenuates nephrosclerosis and improves survival in salt-loaded spontaneously hypertensive stroke-prone rats. J Hypertens 25:1053–1063

Nishizuka Y (1992) Intracellular signaling by hydrolysis of phospholipids and activation of protein kinase C. Science 258:607–614

Noll G, Tschudi M, Nava E, Luscher TF (1997) Endothelium and high blood pressure. Int J Microcirc Clin Exp 17:273–279

Ohanian V, Ohanian J, Shaw L, Scarth S, Parker PJ, Heagerty AM (1996) Identification of protein kinase C isoforms in rat mesenteric small arteries and their possible role in agonist-induced contraction. Circ Res 78:806–812

Panza JA (1997) Endothelial dysfunction in essential hypertension. Clin Cardiol 20:II-26–II-33

Panza JA, Quyyumi AA, Brush JE Jr, Epstein SE (1990) Abnormal endothelium-dependent vascular relaxation in patients with essential hypertension. N Engl J Med 323:22–27

Panza JA, Casino PR, Badar DM, Quyyumi AA (1993a) Effect of increased availability of endothelium-derived nitric oxide precursor on endothelium-dependent vascular relaxation in normal subjects and in patients with essential hypertension. Circulation 87:1475–1481

Panza JA, Casino PR, Kilcoyne CM, Quyyumi AA (1993b) Role of endothelium-derived nitric oxide in the abnormal endothelium-dependent vascular relaxation of patients with essential hypertension. Circulation 87:1468–1474

Panza JA, Garcia CE, Kilcoyne CM, Quyyumi AA, Cannon RO 3rd (1995) Impaired endothelium-dependent vasodilation in patients with essential hypertension. Evidence that nitric oxide abnormality is not localized to a single signal transduction pathway. Circulation 91:1732–1738

Parkington HC, Coleman HA, Tare M (2004) Prostacyclin and endothelium-dependent hyperpolarization. Pharmacol Res 49:509–514

Perry MG, Molero MM, Giulumian AD, Katakam PV, Pollock JS, Pollock DM, Fuchs LC (2001) ET(B) receptor-deficient rats exhibit reduced contraction to ET-1 despite an increase in ET(A) receptors. Am J Physiol Heart Circ Physiol 281:H2680–H2686

Pollock DM, Pollock JS (2001) Evidence for endothelin involvement in the response to high salt. Am J Physiol Renal Physiol 281:F144–F150

Pollock DM, Keith TL, Highsmith RF (1995) Endothelin receptors and calcium signaling. FASEB J 9:1196–1204

Puddu P, Puddu GM, Zaca F, Muscari A (2000) Endothelial dysfunction in hypertension. Acta Cardiol 55:221–232

Raffetto JD, Khalil RA (2008) Matrix metalloproteinases and their inhibitors in vascular remodeling and vascular disease. Biochem Pharmacol 75:346–359

Rembold CM (1990) Modulation of the [Ca2+] sensitivity of myosin phosphorylation in intact swine arterial smooth muscle. J Physiol 429:77–94

Rembold CM, Murphy RA (1988) Myoplasmic [Ca2+] determines myosin phosphorylation in agonist-stimulated swine arterial smooth muscle. Circ Res 63:593–603

Roman RJ, Harder DR (1993) Cellular and ionic signal transduction mechanisms for the mechanical activation of renal arterial vascular smooth muscle. J Am Soc Nephrol 4:986–996

Romero JC, Reckelhoff JF (1999) State-of-the-Art lecture. Role of angiotensin and oxidative stress in essential hypertension. Hypertension 34:943–949

Rosen B, Barg J, Zimlichman R (1999) The effects of angiotensin II, endothelin-1, and protein kinase C inhibitor on DNA synthesis and intracellular calcium mobilization in vascular smooth muscle cells from young normotensive and spontaneously hypertensive rats. Am J Hypertens 12:1243–1251

Ruan X, Arendshorst WJ (1996) Role of protein kinase C in angiotensin II-induced renal vasoconstriction in genetically hypertensive rats. Am J Physiol 270:F945–F952

Ruilope LM, Lahera V, Rodicio JL, Carlos Romero J (1994) Are renal hemodynamics a key factor in the development and maintenance of arterial hypertension in humans? Hypertension 23:3–9

Salamanca DA, Khalil RA (2005) Protein kinase C isoforms as specific targets for modulation of vascular smooth muscle function in hypertension. Biochem Pharmacol 70:1537–1547

Sasajima H, Shima H, Toyoda Y, Kimura K, Yoshikawa A, Hano T, Nishio I (1997) Increased Ca2+ sensitivity of contractile elements via protein kinase C in alpha-toxin permeabilized SMA from young spontaneously hypertensive rats. Cardiovasc Res 36:86–91

Sasaki M, Hori MT, Hino T, Golub MS, Tuck ML (1997) Elevated 12-lipoxygenase activity in the spontaneously hypertensive rat. Am J Hypertens 10:371–378

Sauro MD, Hadden JW (1992) Gamma-interferon corrects aberrant protein kinase C levels and immunosuppression in the spontaneously hypertensive rat. Int J Immunopharmacol 14: 1421–1427

Sauzeau V, Rolli-Derkinderen M, Marionneau C, Loirand G, Pacaud P (2003) RhoA expression is controlled by nitric oxide through cGMP-dependent protein kinase activation. J Biol Chem 278:9472–9480

Schiffrin EL (1994) Intracellular signal transduction for vasoactive peptides in hypertension. Can J Physiol Pharmacol 72:954–962

Schiffrin EL (1995) Endothelin: potential role in hypertension and vascular hypertrophy. Hypertension 25:1135–1143

Schiffrin EL (1996) The endothelium of resistance arteries: physiology and role in hypertension. Prostaglandins Leukot Essent Fatty Acids 54:17–25

Schiffrin EL (1998) Endothelin and endothelin antagonists in hypertension. J Hypertens 16:1891–1895

Schiffrin EL (2001) Role of endothelin-1 in hypertension and vascular disease. Am J Hypertens 14:83S–89S

Schiffrin EL, Touyz RM (1998) Vascular biology of endothelin. J Cardiovasc Pharmacol 32(Suppl 3):S2–S13

Schnackenberg CG, Welch WJ, Wilcox CS (2000) TP receptor-mediated vasoconstriction in microperfused afferent arterioles: roles of O(2)(-) and NO. Am J Physiol Renal Physiol 279:F302–F308

Schroeder AC, Imig JD, LeBlanc EA, Pham BT, Pollock DM, Inscho EW (2000) Endothelin-mediated calcium signaling in preglomerular smooth muscle cells. Hypertension 35:280–286

Schwartzman ML, da Silva JL, Lin F, Nishimura M, Abraham NG (1996) Cytochrome P450 4A expression and arachidonic acid omega-hydroxylation in the kidney of the spontaneously hypertensive rat. Nephron 73:652–663

Shibata R, Morita S, Nagai K, Miyata S, Iwasaki T (1990) Effects of H-7 (protein kinase inhibitor) and phorbol ester on aortic strips from spontaneously hypertensive rats. Eur J Pharmacol 175:261–271

Shimokawa H, Yasutake H, Fujii K, Owada MK, Nakaike R, Fukumoto Y, Takayanagi T, Nagao T, Egashira K, Fujishima M, Takeshita A (1996) The importance of the hyperpolarizing mechanism increases as the vessel size decreases in endothelium-dependent relaxations in rat mesenteric circulation. J Cardiovasc Pharmacol 28:703–711

Skov K, Mulvany MJ (2004) Structure of renal afferent arterioles in the pathogenesis of hypertension. Acta Physiol Scand 181:397–405

Smeda JS, Lee RM, Forrest JB (1988a) Prenatal and postnatal hydralazine treatment does not prevent renal vessel wall thickening in SHR despite the absence of hypertension. Circ Res 63:534–542

Smeda JS, Lee RM, Forrest JB (1988b) Structural and reactivity alterations of the renal vasculature of spontaneously hypertensive rats prior to and during established hypertension. Circ Res 63:518–533

Somlyo AP, Somlyo AV (2000) Signal transduction by G-proteins, rho-kinase and protein phosphatase to smooth muscle and non-muscle myosin II. J Physiol 522(Pt 2):177–185

Spieker LE, Noll G, Ruschitzka FT, Maier W, Luscher TF (2000) Working under pressure: the vascular endothelium in arterial hypertension. J Hum Hypertens 14:617–630

Suematsu E, Resnick M, Morgan KG (1991) Change of Ca2+ requirement for myosin phosphorylation by prostaglandin F2 alpha. Am J Physiol 261:C253–C258

Sugiyama T, Yoshizumi M, Takaku F, Urabe H, Tsukakoshi M, Kasuya T, Yazaki Y (1986) The elevation of the cytoplasmic calcium ions in vascular smooth muscle cells in SHR – measurement of the free calcium ions in single living cells by lasermicrofluorospectrometry. Biochem Biophys Res Commun 141:340–345

Sung BH, Lovallo WR, Teague SM, Pincomb GA, Wilson MF (1993) Cardiac adaptation to increased systemic blood pressure in borderline hypertensive men. Am J Cardiol 72:407–412

Taddei S, Virdis A, Mattei P, Ghiadoni L, Sudano I, Salvetti A (1996) Defective L-arginine-nitric oxide pathway in offspring of essential hypertensive patients. Circulation 94:1298–1303

Taddei S, Virdis A, Ghiadoni L, Salvetti A (1998) Endothelial dysfunction in hypertension: fact or fancy? J Cardiovasc Pharmacol 32(Suppl 3):S41–S47

Taddei S, Virdis A, Ghiadoni L, Salvetti A (2000a) Vascular effects of endothelin-1 in essential hypertension: relationship with cyclooxygenase-derived endothelium-dependent contracting factors and nitric oxide. J Cardiovasc Pharmacol 35:S37–S40

Taddei S, Virdis A, Ghiadoni L, Salvetti G, Salvetti A (2000b) Endothelial dysfunction in hypertension. J Nephrol 13:205–210

Tharaux PL, Chatziantoniou C, Fakhouri F, Dussaule JC (2000) Angiotensin II activates collagen I gene through a mechanism involving the MAP/ER kinase pathway. Hypertension 36:330–336

Thengchaisri N, Hein TW, Wang W, Xu X, Li Z, Fossum TW, Kuo L (2006) Upregulation of arginase by H2O2 impairs endothelium-dependent nitric oxide-mediated dilation of coronary arterioles. Arterioscler Thromb Vasc Biol 26:2035–2042

Tostes RC, Wilde DW, Bendhack LM, Webb RC (1997) Calcium handling by vascular myocytes in hypertension. Braz J Med Biol Res 30:315–323

Touyz RM, Schiffrin EL (1997) Angiotensin II regulates vascular smooth muscle cell pH, contraction, and growth via tyrosine kinase-dependent signaling pathways. Hypertension 30:222–229

Touyz RM, Schiffrin EL (2003) Role of endothelin in human hypertension. Can J Physiol Pharmacol 81:533–541

Touyz RM, El Mabrouk M, He G, Wu XH, Schiffrin EL (1999) Mitogen-activated protein/extracellular signal-regulated kinase inhibition attenuates angiotensin II-mediated signaling and contraction in spontaneously hypertensive rat vascular smooth muscle cells. Circ Res 84:505–515

Turla MB, Webb RC (1987) Enhanced vascular reactivity to protein kinase C activators in genetically hypertensive rats. Hypertension 9:III150–III154

Uehata M, Ishizaki T, Satoh H, Ono T, Kawahara T, Morishita T, Tamakawa H, Yamagami K, Inui J, Maekawa M, Narumiya S (1997) Calcium sensitization of smooth muscle mediated by a Rho-associated protein kinase in hypertension. Nature 389:990–994

Ungvari Z, Csiszar A, Huang A, Kaminski PM, Wolin MS, Koller A (2003) High pressure induces superoxide production in isolated arteries via protein kinase C-dependent activation of NAD(P)H oxidase. Circulation 108:1253–1258

Urakami-Harasawa L, Shimokawa H, Nakashima M, Egashira K, Takeshita A (1997) Importance of endothelium-derived hyperpolarizing factor in human arteries. J Clin Invest 100:2793–2799

Uski T, Andersson KE, Brandt L, Edvinsson L, Ljunggren B (1983) Responses of isolated feline and human cerebral arteries to prostacyclin and some of its metabolites. J Cereb Blood Flow Metab 3:238–245

Vagnes OB, Iversen BM, Arendshorst WJ (2007) Short-term ANG II produces renal vasoconstriction independent of TP receptor activation and TxA2/isoprostane production. Am J Physiol Renal Physiol 293:F860–F867

Van Zwieten PA (1997) Endothelial dysfunction in hypertension. A critical evaluation. Blood Press Suppl 2:67–70

Vanhoutte PM (1992) Role of calcium and endothelium in hypertension, cardiovascular disease, and subsequent vascular events. J Cardiovasc Pharmacol 19(Suppl 3):S6–S10

Vanhoutte PM (1996) Endothelial dysfunction in hypertension. J Hypertens Suppl 14:S83–S93

Vanhoutte PM (2011) Endothelium-dependent contractions in hypertension: when prostacyclin becomes ugly. Hypertension 57:526–531

Vanhoutte PM, Boulanger CM (1995) Endothelium-dependent responses in hypertension. Hypertens Res 18:87–98

Visse R, Nagase H (2003) Matrix metalloproteinases and tissue inhibitors of metalloproteinases: structure, function, and biochemistry. Circ Res 92:827–839

Vyas SJ, Jackson EK (1995) Angiotensin II: enhanced renal responsiveness in young genetically hypertensive rats. J Pharmacol Exp Ther 273:768–777

Wang T, Chan YL (1990) Mechanism of angiotensin II action on proximal tubular transport. J Pharmacol Exp Ther 252:689–695

Watts SW, Rondelli C, Thakali K, Li X, Uhal B, Pervaiz MH, Watson RE, Fink GD (2007) Morphological and biochemical characterization of remodeling in aorta and vena cava of DOCA-salt hypertensive rats. Am J Physiol Heart Circ Physiol 292:H2438–H2448

Widgren BR, Herlitz H, Aurell M, Berglund G, Wikstrand J, Andersson OK (1992) Increased systemic and renal vascular sensitivity to angiotensin II in normotensive men with positive family histories of hypertension. Am J Hypertens 5:167–174

Williams SP, Dorn GW 2nd, Rapoport RM (1994) Prostaglandin I2 mediates contraction and relaxation of vascular smooth muscle. Am J Physiol 267:H796–H803

Xavier FE, Blanco-Rivero J, Ferrer M, Balfagon G (2009) Endothelium modulates vasoconstrictor response to prostaglandin I2 in rat mesenteric resistance arteries: interaction between EP1 and TP receptors. Br J Pharmacol 158:1787–1795

Yaghini FA, Song CY, Lavrentyev EN, Ghafoor HU, Fang XR, Estes AM, Campbell WB, Malik KU (2010) Angiotensin II-induced vascular smooth muscle cell migration and growth are mediated by cytochrome P450 1B1-dependent superoxide generation. Hypertension 55:1461–1467

Yanagisawa M, Kurihara H, Kimura S, Tomobe Y, Kobayashi M, Mitsui Y, Yazaki Y, Goto K, Masaki T (1988) A novel potent vasoconstrictor peptide produced by vascular endothelial cells. Nature 332:411–415

Ying Z, Jin L, Dorrance AM, Webb RC (2004) Increased expression of mRNA for regulator of G protein signaling domain-containing Rho guanine nucleotide exchange factors in aorta from stroke-prone spontaneously hypertensive rats. Am J Hypertens 17:981–985

Yiu SS, Zhao X, Inscho EW, Imig JD (2003) 12-Hydroxyeicosatetraenoic acid participates in angiotensin II afferent arteriolar vasoconstriction by activating L-type calcium channels. J Lipid Res 44:2391–2399

Zervoudaki A, Economou E, Stefanadis C, Pitsavos C, Tsioufis K, Aggeli C, Vasiliadou K, Toutouza M, Toutouzas P (2003) Plasma levels of active extracellular matrix metalloproteinases 2 and 9 in patients with essential hypertension before and after antihypertensive treatment. J Hum Hypertens 17:119–124

Zhao L, Funk CD (2004) Lipoxygenase pathways in atherogenesis. Trends Cardiovasc Med 14:191–195

Zhao H, Joshua IG, Porter JP (2000) Microvascular responses to endothelin in deoxycorticosterone acetate-salt hypertensive rats. Am J Hypertens 13:819–826

Obesity and Diabetes

Maria Angela Guzzardi and Patricia Iozzo

Abstract

The prevalence of chronic, life-threatening diseases is increasing worldwide at an alarming rate. With about 18 million people dying every day, cardiovascular diseases (CVD) play a major role in the all-cause mortality list, prompting health organizations toward the identification of important risk factors and sustainable prevention strategies. Obesity and diabetes represent prominent predisposing conditions for CVD, and their increasing prevalence is becoming an unbearable challenge. In fact, obesity is generally associated with high blood pressure, dyslipidemia, insulin resistance, hyperleptinemia, and a low-grade chronic inflammatory status, which are risk factors of coronary artery disease (CAD), myocardial infarction (MI), congestive heart failure (CHF), and atrial fibrillation (AF). Also, diabetes mellitus and insulin resistance are independently associated with CAD and ventricular dysfunction. More than 1.1 billion of adults worldwide are overweight and around 300 million of them are obese. In addition, around 346 million people suffer from diabetes according to the World Health Organization, which estimates that the 3.4 million of deaths caused by hyperglycemia will duplicate in 2030. Several pathways implied in the development of CVD in obese or diabetic patients have been identified, but therapeutic strategies aimed at reducing mortality showed controversial results. Educational plans aimed at promoting a healthy lifestyle are likely to be crucial to counteract the pandemia of cardiometabolic diseases.

Keywords

Obesity • Diabetes • Cardiovascular disease • Inflammation • Ectopic fat • Fatty acids • Hypoglycemia • Oxidative stress

---

M.A. Guzzardi, PhD (✉) • P. Iozzo, MD, PhD (✉)
Institute of Clinical Physiology, National Research Council (CNR),
Via Moruzzi 1, 56124 Pisa, Italy
e-mail: m.guzzardi@ifc.cnr.it; patricia.iozzo@ifc.cnr.it

2.1 Introduction

The prevalence of chronic, life-threatening diseases is increasing worldwide at an alarming rate. With about 18 million people dying every day, cardiovascular diseases (CVD) play a major role in the all-cause mortality list, prompting health organizations toward the identification of important risk factors and sustainable prevention strategies.

Diabetes and obesity represent prominent predisposing conditions for CVD, and their increasing prevalence is becoming an unbearable challenge.

2.2 Epidemiology

2.2.1 Obesity Pandemic and Cardiovascular Health

Today, more than 1.1 billion adults worldwide are overweight and around 300 million of them are obese. Obesity is also increasing among children, with at least 155 million of individuals affected by overweight or obesity according to the Obesity Task Force (Hossain et al. 2007). Overall, in every country that has adopted a Western lifestyle, including the overconsumption of energy-enriched foods and poor physical activity, obesity has tripled in the last 20 years.

According to the World Health Organization (WHO), overweight and obesity are defined as "abnormal or excessive fat accumulation that may impair health." The body mass index (BMI), a simple body weight-to-height ratio (kg/m^2), is used to classify overweight and obesity in adults, and normal weight is defined as a BMI ranging between 18.8 and 24.9 kg/m^2, whereas BMI cutoff values of 25, 30, 35, and 40 kg/m^2 define overweight, obesity, morbid obesity, and extreme obesity. The BMI does not take into account the distribution of fat masses in the subcutaneous of visceral body compartments, or the relative proportions of fat and lean mass, or gender and age differences. For instance, women have lower muscle and bone mass than men and a greater percentage of subcutaneous than visceral body fat. Older subjects tend to have a higher proportion of fat than younger subjects. There is a wide recognition that central (visceral) obesity plays a prominent role in CVD and other complications of overweight, and therefore the assessment of the waist circumference may be a useful proxy indicator of visceral adiposity to complement the evaluation of a given BMI.

In children up to 19 years old, weight categories are based on percentiles and overweight and obesity cutoff values correspond to the 85th and 95th percentiles of BMI in a sex- and age-matched reference population. No consistent recommendations exist for infants younger than 2 years, in which overweight is often defined as a weight-for-length value ≥95th percentile within a reference population (Ogden et al. 2007).

In the United States, according to The National Health and Nutrition Examination Survey (NHANES), the prevalence of obesity in adults has increased from 15 to 33 %, and that of overweight children has gone from 6 to 19 % between 1980 and

2004. In 2003–2004, >17 % of teenagers were overweight, and 32.9 % of adults were obese, with a 50 % increase in the prevalence of obesity in women during the decade 1994–2004. Ogden et al. (2007) report that the prevalence of overweight increases with age until 80 years, after which it may decline. Moreover, large differences in race-ethnic groups have been shown with a higher prevalence in black women and in Mexican American children and adolescents.

Dramatic changes have occurred also in Europe. National and regional surveys conducted between the late 1980s and 2005 (Berghöfer et al. 2008) showed that obesity ranged between 4 and 28.3 % in men and 6.2 and 36.5 % in women, with geographic heterogeneity and greater prevalences in Italy and Spain. These and other data reported by the WHO Multinational Monitoring of Trends and Determinants in Cardiovascular Disease (MONICA) show trends similar to the ones observed in the USA by the NHANES survey. Similar tendencies have also been reported in the Chinese population, in which the prevalence of obesity has doubled over the last decade (Ding and Malik 2008).

Overweight and obesity are tightly associated with higher blood pressure, dyslipidemia, insulin resistance, hyperleptinemia, and high circulating levels of inflammatory markers, which are independent risk factors for CVD and type 2 diabetes, both in men and in women and increasingly in children.

In fact, obesity is strongly associated with a more frequent development of coronary artery disease (CAD), myocardial infarction (MI), congestive heart failure (CHF), and atrial fibrillation (AF). Sudden death in subjects with severe obesity may be due to the associated dilated cardiomyopathy. Very large studies, including approximately 1.5 million individuals, have repeatedly confirmed that the relationship linking the BMI and CVD mortality has a J shape, in which both underweight and overweight increase mortality (Calle et al. 1999; Berrington de Gonzalez et al. 2010). Smaller studies in patients with heart failure or coronary atherosclerosis have shown the inverse relationship, i.e., that obese patients who suffered from hypertension, CHF, AF, CAD, or MI and revascularization had a lower mortality risk than normal-weight patients. This paradox is explained by the delayed monitoring, older age, and less optimal treatment observed in lean than overweight subjects (Wu et al. 2010; Schenkeveld et al. 2012).

2.2.2 Diabetes Pandemic and Cardiovascular Health

Difficulties in determining the prevalence of diabetes worldwide arise from the use of different diagnostic criteria and the paucity of data, especially in Eastern Europe and Southern Asia, and in younger and older age extremes. The WHO documents that approximately 346 million people worldwide have diabetes. Diabetes mellitus (DM) is characterized by systemic hyperglycemia, insulin resistance at the whole body level and in peripheral organs, including the heart, and a relative or absolute impairment in beta-cell function. Type 2 diabetes mellitus (T2DM), which is related to obesity and lifestyle, accounts for ~90 % of diabetic cases, representing an ominous CVD cause and prognostic factor.

In 2004, Wild et al. (2004) estimated that, given 171 million of people (≥20 years of age) with diabetes in 2000 and assuming that other important risk factors will remain constant, 366 million people will be affected by diabetes in 2030. This is likely to be an underestimate, since risk factors are by themselves growing in frequency and severity. Diabetes prevalence is slightly more prevalent in <60-year-old men and in older women. This might be because there are a greater number of elderly women in most populations, the prevalence of diabetes increases with aging, and the protective effect of estrogens is lost in postmenopausal women (Wild et al. 2004).

Differences in developed (Europe, North America, Japan, Australia, New Zealand) versus developing countries exist. While most diabetic subjects are older than 64 years in developed countries, they are in the 45–64-year age range in developing countries. It has been estimated that developing countries will face the greatest relative increase in the number of people with diabetes by 2030, especially in the Middle East, sub-Saharan Africa, and India.

In 2004 roughly 3.4 million people died from consequences of hyperglycemia, with more than 80 % of diabetes deaths occurring in the low- and middle-income countries. Both WHO and International Diabetes Foundation (IDF) projected a doubling of diabetes deaths between 2005 and 2030.

The increasing prevalence of diabetes is associated with increasing proportion of CVD in the global population. CAD and ventricular dysfunction are main complications affecting diabetic patients. Despite the great advances in the prevention and therapy of CVD, the prognosis in diabetic patients remains poor.

Scholte et al. (2008) showed a vascular phenotype of diffuse coronary artery disease and increased risk of MI in asymptomatic diabetic patients. Moreover, more than one-fourth of patients who experienced MI had comorbid DM, and the percentage goes up to more than 75 % for clinically manifest ischemic heart disease, if including impaired glucose tolerance or impaired fasting glycemia (Lenzen et al. 2006). Similarly, large population studies have shown that the prevalence of CHF is four times higher in diabetic patients than in nondiabetic or prediabetic subjects. In return, the prevalence of DM in patients with CHF ranges between 10 and 30 %, depending on population selection. Indeed, Schramm et al. (2008), studying a large cohort of 3.3 million subjects, showed that patients with DM and without prior cardiovascular events have a risk of MI equivalent to that found in nondiabetic patients with established ischemic heart disease. The available data suggest that the magnitude of hyperglycemia per se might be a predictor of post-MI risk or risk to develop CHF, and the UK Prospective Diabetes Study (UKPDS 35) indicates an increase in the relative risk of 16 % for every 1 % absolute increment in glycosylated hemoglobin (Stratton et al. 2000). DM is an independent predictor of hospitalization and mortality in CHF patients, and CHF with a preserved ejection fraction is a common phenotype among diabetic patients with heart failure. Interestingly, diabetes mellitus in CHF patients with preserved ejection fraction has been related to worst cardiovascular prognosis than CHF with LV systolic dysfunction (Mercer et al. 2012).

Beyond chronic hyperglycemia, the documented increase of cardiovascular risk in diabetic subjects is due to the combined and additive effect of a number of factors, including obesity, dyslipidemia, inflammation, insulin resistance, and hypertension.

All these factors are associated with an increased incidence of CAD and progression of vascular dysfunction, eventually leading to CHF and MI.

Alterations of myocardial structure and function are also promoted by DM in spite of normal coronary arteries and absence of hypertension, as a consequence of multiple mechanisms, mostly related to insulin resistance. This cardiac loss of structure and function is generally referred to as "diabetic cardiomyopathy."

2.3 Pathological Mechanisms

Obesity and diabetes are often considered together when describing risk factors for CVD. This is mostly due to the consensual occurrence of both in many patients, which makes it difficult to disentangle their specific contribution in the development of heart disease. Here we aim at reviewing the mechanisms that have been suggested to underlie a poor cardiovascular prognosis in obese and diabetic patients (Table 2.1).

2.3.1 Hemodynamic Effects

In obesity, the expansion of body mass results into adverse hemodynamic effects. In fact, overweight and obese patients have an increase in total blood and stroke volumes, increasing cardiac output and affecting heart structure and function. Overweight is typically associated with an increase in arterial stiffness and pressure resulting in systemic hypertension. In moderate to severe obesity, the increase in filling pressure and volume may develop into LV chamber dilation (Alpert 2001), increased wall stress, compensatory eccentric hypertrophy, and diastolic dysfunction (Messerli et al. 1987). LV systolic dysfunction might also occur if wall stress persists because of inadequate hypertrophy. Left atria enlargement may follow due to an elevated circulating blood volume and abnormal LV diastolic filling (Lavie et al. 1987), and it may further increase the risk of atrial fibrillation and heart failure.

Overall, cardiovascular structural and hemodynamic changes due to excessive adipose tissue accumulation are associated with congestive heart failure and sudden cardiac death, representing the predominant causes of death in these subjects (Lavie et al. 2009).

In addition, especially if insulin resistance occurs, the tone of small vessels throughout the body, including the brain and heart, may become impaired. Normally, the endothelium supplies vasodilatory nitric oxide (NO) to vascular smooth muscle cells, providing instantaneous adjustments of the vascular tone. Endogenous NO production has cardioprotective effects by dilating coronary arteries in ischemic heart disease and reducing myocardial cell apoptosis and myocardial necrosis. Patients with diabetes have a reduced availability of NO (Dokken 2008). This is due, at least in part, to impairment in insulin signaling, since insulin enhances endothelial nitric oxide synthase-derived NO production through phosphorylation-dependent mechanisms (Yu et al. 2010). Thus, microvascular damage may occur independent of atherosclerosis.

Table 2.1 Pathological mechanisms involved in the development of cardiovascular diseases in obese or diabetic condition

Pathological mechanism	Cardiovascular disease	Experimental model	Overweight and obesity	Diabetes mellitus	References
Hemodynamic overload	Hypertension remodeling	Obese patients	+++	+	Alpert (2001), Dokken (2008), Lavie et al. (1987), Messerli et al. (1987)
Endothelial dysfunction	Hypertension Atherosclerosis Ischemic heart disease	Diabetes and metabolic syndrome patients	++	+++	Dokken (2008), Yu et al. (2010)
Impaired leptin and adiponectin secretion	Atherosclerosis CAD	Obese and diabetic patients, transgenic mice and rats, obese mice (ob/ob; db/db), Zucker rats	+++	++	Enriori et al. (2006), Pischon et al. (2004), Romero-Corral et al. 2008; Schwartz et al. (1996), Szmitko et al. (2007)
Adipocytokines and inflammation	Atherosclerosis CAD	Obese and diabetic patients; transgenic and high-fat diet mice	+++	++	Cinti et al. (2005), Kanda et al. (2006), Shimabukuro (2009)
Oversupply of FFAs: myocardial metabolic switch	Diabetic cardiomyopathy, diastolic and systolic dysfunction	Obese and diabetic patients; high-fat diet mice; obese mice (ob/ob; db/db)	+++	+++	Abel et al. (1999), Belke et al. (2002), Herrero et al. (2006), Iozzo et al. (2002), Peterson et al. (2004), Young et al. (2001)
Myocardial insulin resistance	Diabetic cardiomyopathy: remodeling; diastolic and systolic dysfunction	T1DM and T2DM patients; obese patients; diabetic transgenic mice and rats; high-fat diet mice; obese mice (ob/ob; db/db), Zucker rats	++	+++	Abel et al. (1999), Iozzo et al. (2002), Peterson et al. (2004)
Oversupply of FFAs: DAG and ceramides cardiac accumulation	Diabetic cardiomyopathy, diastolic and systolic dysfunction	Transgenic mice, isolated cardiomyocytes	+++	++	Gudz et al. (1997), Schenk et al. (2008)

Intramyocardial triglyceride accumulation	Diabetic cardiomy-opathy, diastolic and systolic dysfunction	Obese and diabetic patients; transgenic mice	+++	++	Ding and Malik (2008), Dokken (2008)
Epicardial adipose tissue accumulation	Diabetic cardiomy-opathy, diastolic and systolic dysfunction CAD	Obese, glucose intolerant, and diabetic patients	+++	++	Brinkley et al. (2011), Iacobellis and Bianco (2011), Iacobellis et al. (2003), Iozzo (2011), Mazurek et al. (2003), van Marken Lichtenbelt et al. (2009)
Impaired calcium handling	Diabetic cardiomy-opathy, diastolic and systolic dysfunction	Diabetic rat cardiomyocytes	+	++	Belke et al. (2002), Choi et al. (2002), Nobe et al. (1990), Pereira et al. (2006)
Mitochondrial dysfunction and oxidative stress	Diabetic cardiomy-opathy, diastolic and systolic dysfunction	Diabetic patients, diabetic mice (STZ-induced T1D and ob/ob, db/db)	++	+++	Barouch et al. (2003), Boudina et al. (2009), Raha and Robinson (2000), Turko and Murad (2003)
AGEs formation	CAD Diabetic cardiomyopathy	Diabetic patients, diabetic mice	n.d.	+++	Basta et al. (2004), Kilhovd et al. (1999)
Renin-angiotensin axis activation	Hypertension Diabetic cardiomyopathy	Diabetic rats, isolated rat cardiomyocytes	+	+++	Ballard and Schaffer (1996), Ferron et al. (2003), Fiordaliso et al. (2000), Gunasegaram et al. (1999), Privratsky et al. (2003)

Abbreviations: *CAD* coronary artery disease, *FFAs* free fatty acids, *DAG* diacylglycerol, *T1DM* type 1 diabetes mellitus, *AGEs* advanced glycation end products, *STZ* streptozotocin, *n.d.* not defined

2.3.2 Atherosclerosis

Atherosclerosis and CAD are frequent cardiovascular complications in obese and diabetic subjects, accounting for 70 % of deaths in diabetic patients worldwide (Zhang and Chen 2012). The formation of the atherosclerotic lesion within the intima layer of the arterial wall is promoted by inflammatory processes, involving the activation of endothelial cells and smooth muscle cells and leading to increased arterial thickness and stiffness and eventually reduced blood flow to the myocardium.

The atherosclerotic process is initiated by exposure of endothelial cell membranes – lining the inner wall of the vessels – to adhesion molecules that promote the recruitment of circulating monocytes and their maturation into macrophages. The following cascade of events includes further endothelial uptake of lipids, formation of foam cells and of ROS due to lipid peroxidation, and release of cytokines. The progression of the atherosclerotic lesion involves activation of subtending smooth muscles cells that migrate from media to intima and proliferate and promote synthesis of extracellular matrix molecules (i.e., collagen and proteoglycans), fibrosis, and plaque formation. Eventually, a hypoxic core can form within the plaque, which leads to cellular necrosis and apoptosis and accumulation of lipid and toxic molecules. The rupture of the fibrous cap of the plaque promotes the formation of a thrombus and obstruction of coronary vessels, preventing oxygen and nutrients from reaching the myocardium and leading to tissue death.

In obese and diabetic subjects, hypertension and endothelial dysfunction increase the susceptibility to atherosclerotic-initiating injuries. In both diseases, dyslipidemia is characterized by increased circulating levels of free fatty acids (FFAs), triglycerides, cholesterol, and low-density lipoproteins (LDL), which correlate with the release of proinflammatory and proatherogenic cytokines (i.e., tumor necrosis factor-α, TNF-α, and interleukin-6, IL-6), lipid peroxidation products, and reactive oxygen species (ROS), and the development of low-grade systemic inflammation. These factors promote the activation of endothelial cells and the progression of atherosclerosis. Dyslipidemia is addressed more extensively in a different chapter of this book.

In the following paragraphs, we focus on the mechanisms whereby adipose tissue dysfunction and hyperglycemia increase the risk of CAD and heart failure (Fig. 2.1).

2.3.3 Adipokines and Inflammation

Far from being a simple energy-storing organ, adipose tissue is now recognized as an endocrine organ able to secrete cytokines and hormones, beyond FFAs. These hormones and cytokines can act in an autocrine, paracrine, or endocrine manner. They have inflammatory, proatherogenic, and anti-inflammatory properties, and they control food intake and energy metabolism.

Leptin and adiponectin are cytokines produced by adipose tissue and exert a protective role by central and peripheral energy homeostasis. By interacting with

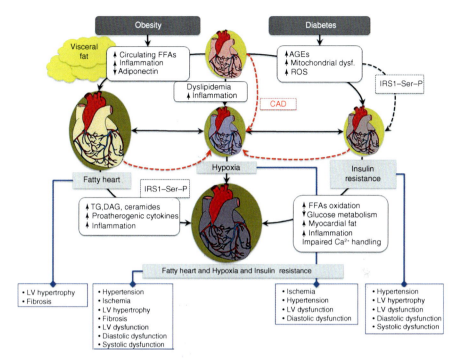

Fig. 2.1 Suggested mechanisms underlying the development of cardiovascular disease in obesity and diabetes. Two examples of cascades are shown, as primarily related with adiposity in obesity (*left*) or hyperglycemia and insulin resistance in T2DM (*right*). However, profound overlaps exist in the mechanisms and outcomes, given their *vicious cycle* nature and because overweight and impaired glucose tolerance or T2DM are usually coexisting. *Abbreviations*: *FFAs* free fatty acids, *AGEs* advance glycation end products, *ROS* reactive oxygen species, *CAD* coronary artery disease, *IRS1-Ser-P* insulin-receptor substrate-serine-phosphate, *TG* triglycerides, *DAG* diacylglycerol, *LV* left ventricle

the hypothalamus-pituitary-adrenal axis, leptin suppresses appetite and enhances energy expenditure, thus reducing ectopic fat accumulation. Animal models that are genetically deficient in leptin (ob/ob mouse) or leptin receptor (db/db mouse and Zucker rat) develop obesity, dyslipidemia, and insulin resistance (Schwartz et al. 1996), which can be reversed by the exogenous administration of leptin. Most obese and many diabetic subjects do not have low but rather elevated circulating leptin, suggesting that they may be resistant or tolerant to the action of this hormone (Enriori et al. 2006). Notably, high circulating leptin concentrations have been suggested to exert a proinflammatory role and have been associated with increased coronary artery atherosclerosis and calcification. In fact, a positive correlation between leptin and C-reactive protein (CRP) has been shown. Moreover, higher levels of leptin have been found in patients with CAD, correlating with the severity of the disease (Romero-Corral et al. 2008).

Adiponectin is a crucial mediator in peroxisome proliferator-activated receptor gamma (PPAR-γ) signaling. The transcription factor PPAR-γ regulates glucose and fatty acid uptake and oxidation in liver, skeletal muscle, and myocardium. *In vitro* studies have shown that adiponectin has insulin-sensitizing, anti-inflammatory, and atheroprotective effects. Adiponectin-deficient mice have an upregulation of endothelial adhesion molecules and increased cell proliferation in response to vascular injury. Arterial wall neointima thickening and enhanced thrombus formation and platelet aggregation have also been shown in these animals as compared with wild-type mice (Matsuda et al. 2002). Adiponectin administration reverses the proatherosclerotic state and attenuates cell proliferation. It decreases the activity of macrophage scavenger receptors, which are responsible for lipid accumulation and foam cell formation, and suppresses the proliferation, migration, and calcification of these cells in the atherosclerotic lesion. As an anti-inflammatory molecule, adiponectin reduces the release of CRP and interleukin-8 and endothelial cell apoptosis (Ouedraogo et al. 2007). Besides its role in atherosclerotic disease, adiponectin lowers blood glucose and FFA levels. In humans, reduced plasma adiponectin or some polymorphisms in the adiponectin gene have been associated with obesity, development of T2DM, and increased risk of cardiovascular diseases (Szmitko et al. 2007). Adiponectin plasma levels have been negatively correlated with CRP, development of CAD, and increased carotid artery intima-media thickness (CIMT) and atheroma burden. In spite of the above observations, some controversial findings have been reported on the association between adiponectin and cardiovascular risk in clinical trials. In several studies, high adiponectin levels were associated with a lower risk of myocardial infarction independently of diabetes, hypertension, BMI, and family history of myocardial infarction (Pischon et al. 2004). Conversely, the Health Study (Lawlor et al. 2005) and the Strong Heart Study (Lindsay et al. 2005) have failed to identify adiponectin as an independent predictor of cardiovascular disease. The presence of comorbidities or gender effects may to some extent explain the discrepancies, since studies performed in men tend to find a positive correlation between low adiponectin and high CVD risk. Other adipokines, such as retinol-binding protein 4, resistin, visfatin, and omentin, are involved in the regulation of insulin sensitivity, substrate oxidation, and inflammation, even though their role in the development of CVD requires additional investigation.

Therefore, adipokines represent key mediators in the adipose tissue-myocardium axis, which plays a prominent role in the cardiovascular health of obese and overweight subjects. It is believed that the enlargement of adipocytes leads to local hypoxia, promoting the production of free radicals, cell apoptosis, and release of local chemokines, such as monocytes chemoattractant protein-1 (MCP-1) that result in a low-grade systemic inflammation; inhibiting adiponectin gene expression; and interfering with insulin signaling, thereby stimulating lipolysis and the release of FFAs and reducing glucose uptake. Mice overexpressing MCP-1 develop insulin resistance (Kanda et al. 2006) and exhibit higher levels of TNF-α and IL-6. Increased inflammation and newly recruited macrophages forming the typical crown-like structure around dead adipocytes have been observed in obese subjects (Cinti et al. 2005). Visceral adipose tissue (VAT) is considered more dangerous than subcutaneous fat in terms of cardiovascular risk. It appears more prone to a proinflammatory endocrine and metabolic asset, which has been also involved in T2DM onset (Shimabukuro 2009).

Besides the systemic effect, circulating inflammatory cytokines play a direct role in the metabolic regulation of the myocardium, by inhibiting AMP-activated kinase (AMPK), which reduces glucose metabolism, which is a required oxygen-sparing substrate in ischemia. AMPK is a sensor and modulator of energy metabolism, and it is activated when the AMP:ATP ratio increases (e.g., in ischemic conditions). AMPK can act by inducing both myocardial fatty acid mitochondrial transport and oxidation via the reduction of acetyl-CoA carboxylase and malonyl-CoA levels and increase in carnitine palmitoyltransferase-1 activity. AMPK also promotes myocardial GLUT4 expression and translocation to the sarcoplasma. In conditions of oxygen deficiency, AMPK acts providing energy from non-oxidative pathways (i.e., glycolysis).

Overall, a tight connection between inflamed adipose tissue and the development of insulin resistance, both systemic and myocardial, has been suggested (Iozzo 2009). Adipose tissue dysfunction, which is common in obesity and T2DM, may aliment the progression of atherosclerosis and the damage caused by ischemia.

2.3.4 Oversupply of FFA and Myocardial Metabolism

Under physiological conditions, 60–80 % of the energy required by the heart is provided by lipid oxidation, while glucose oxidation accounts for most of the remainder, and only a small part is provided by lactate, ketones, and amino acids. Therefore, FFAs represent the main fuel of cardiomyocytes, given that FFA breakdown gives more energy than any other substrates. As an example, the oxidation of palmitic acid produces 129 ATP per molecule, while one molecule of glucose produces only 36 ATP. FFAs enter myocardial cells mainly by fatty acid transport proteins associated to the plasma membrane (FATPs) and FA translocase (FAT/CD36), whose expression is regulated by circulating FFAs levels.

Glucose metabolism is also important to maintain normal cardiac function, being at least fourfold greater in the heart than in skeletal muscle or adipose tissue. It is associated with lower oxygen consumption and contributes to keep ATP production constant and appropriate to the heart needs. Glucose transport into myocytes occurs through glucose transporter proteins GLUT4, the major heart insulin-responsive glucose transporter, and GLUT1. In physiological condition, myocardial FFAs and glucose metabolism are modulated according to their blood levels (Hue and Taegtmeyer 2009).

Obesity and T2DM are associated with greater circulating FFA levels reaching the heart. The augmented FFA availability may promote β-oxidation and inhibit glucose oxidation. Therefore, myocardial insulin resistance (MIR) may develop as a consequence of increased blood FFA levels, according to Randle's glucose-FA cycle, as initially described in skeletal muscle and subsequently observed in the myocardium. In addition, the activation of the PPARα/PGC1 transcriptional pathway by FFAs stimulates the transcription of genes involved in FA uptake (FATP1 and CD36, mitochondrial malonyl-CoA decarboxylase) and β-oxidation (medium- and long-chain acyl CoA dehydrogenase and hydroxyacyl CoA dehydrogenase).

At the same time, PPARα activation inhibits gene expression of pyruvate dehydrogenase kinase 4, reducing glucose oxidation. Through this "feed-forward" mechanism, the heart attempts to adapt to a high-fat environment, enhancing FA utilization and inducing triglyceride accumulation, at the expense of glucose uptake and metabolism. This condition aggravates MIR.

Metabolic changes in the myocardium can affect the function of the heart, as observed in animal models and in obese and diabetic human subjects.

Studies in mice have shown that the heart-specific ablation of GLUT4 (Abel et al. 1999) or of insulin receptors (Belke et al. 2002) alters the morphology of the heart and provokes ischemia-associated diastolic and systolic dysfunction or a global impairment in cardiac function. In hypertrophic hearts adapted to an impaired metabolism, PPARα is activated and results in contractile dysfunction (Young et al. 2001).

Positron emission tomography (PET) studies have shown that systemic IR in T2DM patients is associated with MIR (Iozzo et al. 2002). Moreover, increased myocardial FA oxidation and decreased glucose utilization were shown in T1DM patients (Herrero et al. 2006) and in obese subjects (Peterson et al. 2004), and they were associated with a decrease in cardiac efficiency. These cardiometabolic alterations likely contribute to the pathogenesis of ventricular dysfunction associated with obesity and diabetes (Iozzo et al. 2002).

The association between MIR and heart dysfunction can be explained in terms of cardiac efficiency. Oxidation of FA requires more oxygen compared to glucose oxidation, and an increased oxygen consumption (MVO2) is associated with a reduction in cardiac efficiency, since more oxygen is used to perform a given amount of cardiac work, in obese and insulin-resistant subjects (Peterson et al. 2004). The oversupply of FFA to the heart is also associated with mitochondrial dysfunction, oxidative stress, and other mechanisms that reduce cardiomyocyte oxygen availability. As a result, the insulin-resistant, diabetic heart is particularly vulnerable to ischemia/reperfusion, in which the coupling between oxygen consumption and ATP production is crucial.

2.3.5 Oversupply of FFA and Intramyocardial Triglycerides and Lipid Intermediates

Once FFA uptake exceeds the oxidative capacity of the heart, they are diverted to non-oxidative pathways, leading to intracellular triglyceride accumulation and the biosynthesis of lipid intermediates such as ceramides and diacylglycerides (DAG).

Intramyocardial triglyceride content is generally increased from two- to fourfold in obese and T2DM patients with respect to controls. In these patients, the increase in cardiac fat is associated with that in LV mass and work and with an impaired diastolic function (Kankaanpää et al. 2006). However, the role of triglyceride accumulation in the pathogenesis of diastolic dysfunction is controversial because it is not consistently reproduced in animal and clinical studies. In fact, the transgenic overexpression of the triglyceride synthesizing enzyme diacylglycerol acetyltransferase

1 (DGAT1) resulted in a protective function even in the presence of heart hypertrophy. In addition, multivariable analysis models could not find a robust association between cardiac triglyceride accumulation and early (E) and late (A) ventricular filling velocity (E/A) in diabetic patients (McGavock et al. 2007).

The lipid-derived metabolites ceramide and DAG, instead, clearly exert a toxic effect by interfering with insulin signaling through serine phosphorylation of the insulin-receptor substrate (IRS1). They also exert a lipotoxic effect by locally promoting inflammation, apoptosis, and oxidative stress (Guzzardi and Iozzo 2011). In isolated cardiomyocytes, ceramides inhibit the mitochondrial respiratory chain by interacting with complex III (Gudz et al. 1997); activate caspase-3-like protease, which eventually results in cell apoptosis; and can inhibit the Akt/PKB signaling involved in glucose and protein metabolism and in cell survival pathways.

DAGs are intermediate metabolites forming during triglyceride synthesis or hydrolysis. They can activate serine/threonine-specific protein kinases belonging to the superfamily of mitogen-activated protein kinases (MAPKs), such as IκB and protein kinase C. MAPKs regulate a complex signaling network interfering with several cellular processes, including apoptosis, inflammation, T-cell differentiation, and insulin action (Schenk et al. 2008).

The specific contribution of intramyocardial lipid accumulation to the pathogenesis of obesity or diabetic cardiomyopathy is mostly associated to the development of cardiac inflammation and insulin resistance.

2.3.6 Epicardial Adipose Tissue

Epicardial adipose tissue (EAT) is located between the myocardium and the visceral pericardium, different from pericardial fat, which is situated on the external surface of the parietal pericardium. No muscle fascia divides myocardium and EAT, and EAT is nourished by branches of coronary arteries, which are surrounded by this fat depot. Thus, the endocrine function and physical mass can locally reflect on myocardial structure and function (Iacobellis and Bianco 2011). EAT adipocytes are generally smaller and contain a larger number of preadipocytes with respect to other visceral fat depots, sharing the same embryological origin from the mesoderm. Moreover, lower glucose utilization rates and higher rates of FFA uptake, incorporation into lipid, oxidation, and release have been shown in EAT with respect to other fat tissues.

A protective and physiological function has been attributed to EAT, as it may serve as a buffer, buffering the excess of circulating FFAs or providing substrates to the heart when needed. Moreover, a supportive mechanical function and thermoregulatory role have also been proposed, since the expression of uncoupling protein-1 (UCP-1) and upregulation upon exposure to cold air were found (van Marken Lichtenbelt et al. 2009).

Obesity, visceral fat volume, and waist circumference are correlated with EAT thickness and volume, which also correlates to the number of metabolic syndrome components, with a progressive increment from lean to obese subjects with normal glucose tolerance to those with impaired glucose tolerance and T2DM (Iozzo 2011;

Iacobellis et al. 2003). Moreover, EAT volume has been positively associated with left ventricle hypertrophy, with altered cardiovascular metabolism and inflammation, and with the progression of CAD and negatively associated with heart failure in patients with left ventricular systolic dysfunction.

In patients suffering from severe CAD, thickened connective tissue septa, inflammatory cell infiltrates, increased reactive oxygen species, and expression of inflammatory cytokines have been found within epicardial adipose tissue (Mazurek et al. 2003), prompting clinicians' and researchers' attention to its detrimental role in CVD, with particular regard to atherosclerosis and cardiac metabolic regulation.

In the Multi-Ethnic Study of Atherosclerosis, EAT thickness was associated with carotid stiffness but not with intima-media thickness after correction for cardiovascular risk factors (Brinkley et al. 2011). Moreover, imaging studies showed that patients with CAD have increased EAT volume compared to controls. However, the association between the volume or thickness of epicardial fat and the severity of CAD is elusive; in fact, only in some studies a progressive association between the amount of fat and the number of stenotic vessels or total coronary obstruction or stable versus unstable angina or vascular calcification was found (Iozzo 2011).

Studies on animal models and clinical trials suggest a number of putative mechanisms by which EAT might play a significant role in the inflammatory process within the atherosclerotic plaque. The process can involve innate immune cells through activation of Toll-like receptors (TLRs) and nuclear translocation of the proinflammatory transcription factor NF-κB. In fact, a lower expression of adiponectin was found in epicardial fat from CAD patients compared to subjects without CAD, in favor of the expression of detrimental cytokines, such as TNF-α, IL-6 and others (Mazurek et al. 2003), which may reach the myocardium and the coronary wall by direct diffusion or through the adventitial circulation. EAT growth also induces cell surface expression of adhesion molecules, such as MCP-1 and intercellular adhesion molecules (ICAM), which enhance adhesion of monocytes to endothelial cells and promote foam cell formation, smooth muscle cell proliferation, and migration and plaque destabilization. The propagation of inflammation to the underlying arterial wall may alter the balance between nitric oxide, endothelin 1, and superoxide production, promoting vasoconstriction. Increased expression of secretory type II phospholipase A_2 (sPLA_2-IIA), an enzyme involved in retention of LDL in the subendothelial space (Iacobellis and Bianco 2011), and higher levels of oxidative stress associated with the down regulation GLUT4 have also been found in EAT with respect to subcutaneous fat in patients with CAD.

Overall, EAT may have the protective or pathogenic functions observed in abdominal visceral adipose tissue, buffering the excess of lipid substrates, especially under high FFA conditions, and modulating substrate fluxes to the heart, but it can also interact with the control of vascular tone and blood flow when inflammation occurs, affecting atherogenesis and vascular remodeling, blood pressure, myocardial hypertrophy, and adipogenesis through paracrine or vasocrine (vasa vasorum) mechanisms.

In addition, increased EAT thickness in obese subjects is correlated with the intramyocardial fat content, which may reflect "spillover" of FFA from EAT. The two depots may concur to increase the accumulation of toxic lipid intermediate species.

2.3.7 Impaired Calcium Handling

Myocardial contractile function is regulated by intracellular calcium (Ca^{2+}) concentrations. In response to an action potential, a Ca^{2+} inward current occurs through the L-type Ca^{2+} channel. This initiates the Ca^{2+}-induced calcium release mechanism from sarcoplasmic reticulum (SR) via activation of the ryanodine receptor and reduction of sarcolemma sodium/calcium exchanger process to increase intracellular Ca^{2+}. Then, diastolic relaxation occurs, resulting from a decline in intracellular Ca^{2+} levels through activation of the SR Ca^{2+} pump (SERCA2a) and sarcolemma Ca^{2+} ATPase and restoration of the sodium/calcium exchanger process.

In physiological conditions, insulin promotes both cardiomyocyte contraction and relaxation by stimulating sarcoplasmic Ca^{2+} inflow via L-type channel, the expression of ryanodine receptor, and SERCA2a or by reversing the sodium/calcium exchanger.

Myocardial insulin resistance has been associated with perturbations in calcium handling in both *in vivo* and *in vitro* models. In type 1 diabetic rodents, decreased activities of the ryanodine receptor and SR Ca^{2+} ATPase (Choi et al. 2002; Pereira et al. 2006) have been observed and associated with reduced SR calcium stores and Ca^{2+} efflux by sodium/calcium exchanger. Type 2 diabetic rodents also showed lower expression of ryanodine receptor and SR calcium load (Belke et al. 2004).

Consistently, voltage-clamp studies from isolated diabetic hearts also documented that a prolonged action potential may occur (Nobe et al. 1990). In fact, depressed transient outward currents and delayed repolarization have been observed in isolated diabetic cardiomyocytes. Overall, impaired calcium homeostasis results in decreased myocardial contractility and relaxation. Therefore, a prolonged elevation in intracellular calcium concentration may be responsible for the cardiac diastolic dysfunction in diabetic cardiomyopathy.

2.3.8 Mitochondrial Dysfunction and Oxidative Stress

Increase of ROS production is a hallmark of the diabetic heart and contributes to develop and exacerbate diabetic cardiomyopathy. In the myocardium, 90 % of ROS derive from mitochondria (Raha and Robinson 2000), resulting from an imbalance between increased ROS generation and reduced antioxidant mechanisms. In obese and diabetic subjects, augmented ROS generation in the heart is mainly related to increased β-oxidation, altered mitochondrial expression of electron transport chain and oxidative phosphorylation components, mitochondrial dysfunction, and other extramitochondrial mechanisms related to the accumulation of lipid intermediates. Mitochondrial dysfunction translates into reduced oxidative capacity and ATP production and increased generation of ROS. The delivery of reducing equivalents to the electron transport chain increases in proportion to mitochondrial β-oxidation. Once the oxidative phosphorylation components become insufficient to consume all of them, an increase of superoxide production occurs.

In fact, proteomic analysis of hearts from streptozotocin-induced diabetic mice and obese type 2 diabetic mice has shown that in these hearts, there is a reduced expression of mitochondrial proteins involved in the electron transport chain, creatine kinase, voltage-dependent anion channel 1 (Turko and Murad 2003), and oxidative phosphorylation components (Boudina et al. 2005) and increased levels of β-oxidation proteins.

An elevated production of ROS may be responsible for cellular damage through several mechanisms. ROS promote lipid oxidation to form lipid peroxidation products, which activate caspase signaling in ob/ob and db/db hearts enhancing cell apoptosis (Barouch et al. 2003). Moreover, they interfere with DNA repair pathways leading to DNA damage and cell aging in diabetic animals.

By promoting cellular injury and death, ROS contribute to the cardiac remodeling and functional alterations of diabetic cardiomyopathy. Moreover, oxidative stress is closely related to inflammation, and circulating ROS might foster the activation of the immune system and macrophage recruitment, affecting the atherosclerotic process.

Diabetes and obesity-related mitochondrial alterations involve the activation of uncoupling proteins (UCPs). UCPs are mitochondrial inner membrane proteins, which regulate the mitochondrial membrane potential by dissipating the proton gradient. They transport protons from the intermembrane space to the mitochondrial matrix, thus uncoupling the electron transport chain from oxidative phosphorylation and reducing mitochondrial ATP synthesis. UCP activity is enhanced in diabetic hearts and in *in vitro* cardiac fibers with a specific deletion of the insulin receptor and treated with FAs (Boudina et al. 2009). Thus, uncoupling might be responsible for ROS accumulation, due to the inability of mitochondria to reduce superoxide, and to the reduced cardiac efficiency observed in obese and diabetic hearts. Nevertheless, mitochondrial uncoupling in these models might also be a ROS-mediated mechanism activated to reduce the membrane potential and limit further ROS accumulation, thereby representing an adaptive response against an excessive FA uptake and oxidation.

Therefore, it is still on debate whether mitochondrial dysfunction plays a role in the pathogenesis of myocardial insulin resistance or can be considered as an adaptive consequence in the diabetic heart.

2.3.9 Advanced Glycation End Products

Hyperglycemia is considered one of the prominent causes of diabetes complications, and its effect is attributed in part to the formation of advance glycation end products (AGEs). AGEs are generated constantly in the body, but their production is accelerated in diabetes due to sugar availability. Elevated levels of AGEs have been reported in T2DM patients with CAD (Kilhovd et al. 1999).

AGEs are a heterogeneous group of compounds derived by nonenzymatic reactions of reducing sugars with free amino groups of proteins, lipids, and nucleic acids. They exert a detrimental function, by covalently binding to proteins, which alters their structure and function or by interacting with a class of cell surface AGE-binding receptors, called RAGE. The activation of the AGE-RAGE axis has been associated with several events associated with diabetic complications. Animal and human studies have shown that AGEs significantly contribute to the initiation and progression of atherosclerosis, by accumulating in vascular walls, where they alter the extracellular matrix permeability and the antithrombotic proprieties of the endothelium and promote the release of inflammatory cytokines and macrophage activation and the expression of adhesion molecules and chemokines (Basta et al. 2004). Increased AGE levels promote the oxidation and deposition of LDL cholesterol in the vessel wall, resulting in arterial stiffening and progression of atherosclerosis. Therefore, their accumulation causes alterations of the vascular tone in both coronary and other arteries and is associated with CAD and diabetic microvascular complications (i.e., retinopathy and nephropathy).

2.3.10 Activation of the Renin-Angiotensin System

The activation of the renin-angiotensin system (RAS) in diabetic subjects is a known and documented mechanism. By promoting sodium retention and increasing the systemic blood volume, it contributes to hypertension and vascular diabetic complication, including nephropathy and CVD. Hyperglycemia is thought to be the primary cause of renal RAS activation, acting on proximal tubule salt reabsorption. However, the nature of RAS activation in diabetic patients is controversial, since it is difficult to isolate the direct effect of hyperglycemia from other systemic and intrarenal factors that can activate RAS.

Besides renal RAS activation, it has been shown that cardiac activation of RAS plays a role in the development of cardiomyopathy in diabetic subjects through autocrine and paracrine actions. In fact, the peptide angiotensin II (AngII), which is generated in cardiomyocytes and cardiac fibroblast in response to high glucose exposure, may promote cardiac myocyte hypertrophy, fibroblast proliferation, collagen deposition, and cell death pathways (Fiordaliso et al. 2000). Experimental data on animal models reported that RAS activation induces myocyte and endothelial cell oxidative stress and apoptosis and myocardial ischemia by increasing the oxidative activity of NADPH (Privratsky et al. 2003), ROS production, and calcium overload, as mediated by a reduced activity of sodium/calcium and sodium/hydrogen exchangers (Ballard and Schaffer 1996; Gunasegaram et al. 1999) and an activation of T-type calcium channels (Ferron et al. 2003). Moreover, cardiac AngII interferes with insulin signaling by promoting IRS1 tyrosine phosphorylation.

2.3.11 Therapy of Obesity and T2DM: Controversies in the Cardiovascular Outcome

Intentional weight loss in obesity, especially if associated with moderate physical activity, and glucose control in diabetes result in a reduction of the associated health burden (Chugh and Sharma 2012). The association of unintentional weight loss and mortality and the one between intensive glucose control and macrovascular disease are more complex.

In obese individuals, including patients with preexisting CVD, the recent Sibutramine Cardiovascular OUTcomes trial (SCOUT) has shown that even modest intentional weight loss over short-term (6 weeks) and longer-term (6–12 months) periods is associated with a reduction in cardiovascular mortality in the subsequent 4–5 years (Caterson et al. 2012). In some previous studies, the relationship between weight loss and mortality has been counterintuitive. Adverse changes may concur, including the reduction in muscle mass or the psychological effects of dietary deprivation or the increment in unfavorable habits (e.g., smoking), all of which have been shown in association with weight loss (Berentzen and Sorensen 2006). In addition, attention has to be paid at distinguishing wanted from unintentional weight loss, especially in elderly people, since unwanted weight loss may be associated with an underlying disease or malnutrition state and has a negative prognostic significance (Yaari and Goldbourt 1998). Moreover, caution is required in the use of medications to modify the delicate balance of hedonic and homeostatic factors controlling food intake and body weight, as witnessed by the recent withdrawal of drugs (sibutramine and rimonabant) in relation with their cardiovascular or central side effects. Several compounds are in development, acting centrally either to limit food intake (targeting neuropeptide Y, AgRP, MCH1 receptors, leptin sensitizers) and gastrointestinal fat absorption (enzyme inhibitors) or to increase energy expenditure or reduce adipose tissue expandability. Bariatric surgery leads to remarkable weight loss, but it cannot be generalized. Instead, it may serve as an investigational model to understand the role played by the gut and its neural and hormonal outputs in reversing obesity-related diseases, supporting the development of drugs that mimic the effects of bariatric surgery.

In diabetes, the intensive use of medications to achieve an optimal glucose control has shown benefits over the progression of microvascular but not macrovascular complications or mortality (ADVANCE Collaborative Group et al. 2008). The UK Prospective Diabetes Study (UKPDS) was started in 1977 and was designed to evaluate if the intensive use of insulin or sulfonylureas to achieve fasting plasma glucose levels <6 mmol/L could alleviate the risk of micro- and macrovascular complications, as compared with the conventional treatment of T2DM (at that time consisting in diet alone with the addition of medications if fasting plasma glucose was ≥15 mmol/L). It included 3,867 subjects, 70 % of whom were assigned to the intensive treatment arm, leading to a 25 % risk reduction in microvascular end points and a borderline reduction in rates of myocardial infarction in the 10-year follow-up (UKPDS 1998). The Action in Diabetes and Vascular Disease (ADVANCE) trial includes 11,140 patients recruited between 2001 and 2003 and followed for a median period of 5 years. There were no significant effects of the type of glucose control on major macrovascular events and death from cardiovascular causes ($p=0.12$) (ADVANCE Collaborative Group et al. 2008). The Action to Control the Cardiovascular Risk in Diabetes (ACCORD) study, comparing the effect of intensive versus standard therapy, also addressing an intensive versus standard blood pressure or lipid control, on the composite outcome of myocardial infarction, stroke, or cardiovascular death ended its intensive therapy after 3.5 years because of a significant increase in deaths in this group (ACCORD Study Group et al. 2007). The combination of fenofibrate and simvastatin did not improve the cardiovascular risk, and the intensive control of blood pressure did not reduce the composite rate of fatal and nonfatal major cardiovascular events, but reduced the annual rates of stroke (Action to Control Cardiovascular Risk in Diabetes Study Group et al. 2008; ACCORD Study Group et al. 2010a, b; Bonds et al. 2012). Finally, the Veterans Affairs Diabetes Trial (VADT), including 1,791 patients recruited in 2000–2003 and followed up until 2008, was designed to specifically address the effects of intensive and standard glucose control (i.e., an absolute difference of 1.5 % in HbA1c) on cardiovascular events in military veterans (Duckworth et al. 2009). There were no significant differences in cardiovascular events or deaths or time to death between the two treatment arms, but a borderline increase in sudden death was seen, and the occurrence of adverse events, especially dyspnea, was more frequent in the intensive group.

In the above studies, the intensive approach was associated with weight gain and more frequent hypoglycemic episodes, especially in patients who were treated with insulin. For example, in the ACCORD trial, the difference between the two treatment arms was three-fold, and hypoglycemia required emergency medical care in a majority of cases and led to loss of consciousness in 25 % of the events. Lower food intake appeared to be an antecedent factor in approximately half of the episodes. Animal studies conducted in the 1970s have clearly documented that acute hypoglycemia deteriorates the severity and the histological extension of preexisting myocardial ischemia (Libby et al. 1975). Hypoglycemia triggers a counterregulatory cascade, including the activation of the sympathetic nervous system. It is of note that insulin is also a potent suppressor of lipolysis, resulting in a powerful reduction in the levels of circulating FFAs, which are the main fuel of the heart. In recent studies, we have shown that the suppression of FFA levels by using a nicotinic acid analogue can alter cardiac function. Therefore, cardiac substrate deprivation and a high sympathetic stimulation occurring during hyperinsulinemic hypoglycemia may counterbalance the beneficial cardiovascular effect of an optimal glucose control and the positive effects of insulin on myocardial metabolism, perfusion, and function. Together with hypoglycemia, the duration and severity of hyperglycemia and of cardiovascular disease prior to the initiation of the trials have been suggested to explain the negative or neutral outcomes of the above trials.

2.4 Summary and Perspectives

Obesity and diabetes are reaching pandemic proportion worldwide, and their increasing prevalence is associated with an increasing incidence of cardiovascular disease. Moreover, the growing diffusion of overweight and obesity among children and adolescent is becoming alarming and will lead to a proportional rise of adults affected by metabolic and cardiovascular disease in the next future.

Several mechanisms have been illustrated in this chapter, which contribute to the complex pathogenesis of cardiovascular disease in obese and diabetic subjects, mostly involving high levels of circulating fatty acids and other proinflammatory molecules, the impairment of mitochondrial activity, and buildup of oxidative stress.

Intentional weight loss and the use of drugs for glycemic and lipid or blood pressure control have shown beneficial effects in reducing cardiovascular mortality in some, but not all, clinical trials. However, the number and complexity of pathways involved in whole body metabolic regulation, some unexpected side effects of treatment and cardiac responses, together with multiple lifestyle-related or inherited factors and the interference of aging, may account for the controversial results concerning the ways to address cardiovascular health in patients with obesity and diabetes.

Further clinical and preclinical research is required to identify primary mechanisms responsible for the onset of vascular and myocardial dysfunction as a consequence of excessive fat accumulation, hyperglycemia, and hyperinsulinemia. We need to understand why antidiabetic and weight loss interventions are not always beneficial by carefully addressing the counteracting effects of hypoglycemic episodes and unwanted weight loss. Projects aimed at promoting public health awareness should parallel and would optimize the outcome of research findings. In particular, the education of children, who are very receptive, on the prevention of diseases through a healthy lifestyle, including nutrition, physical activity, alcohol, and smoking-oriented programs, would make a difference in the health of future generations.

References

Abel ED, Kaulbach HC, Tian R, Hopkins JC, Duffy J, Doetschman T, Minnemann T, Boers ME, Hadro E, Oberste-Berghaus C, Quist W, Lowell BB, Ingwall JS, Kahn BB (1999) Cardiac hypertrophy with preserved contractile function after selective deletion of GLUT4 from the heart. J Clin Invest 104(12):1703–1714

ACCORD Study Group, Buse JB, Bigger JT, Byington RP, Cooper LS, Cushman WC, Friedewald WT, Genuth S, Gerstein HC, Ginsberg HN, Goff DC Jr, Grimm RH Jr, Margolis KL, Probstfield JL, Simons-Morton DG, Sullivan MD (2007) Action to Control Cardiovascular Risk in Diabetes (ACCORD) trial: design and methods. Am J Cardiol 99(12A):21i–33i

ACCORD Study Group, Cushman WC, Evans GW, Byington RP, Goff DC Jr, Grimm RH Jr, Cutler JA, Simons-Morton DG, Basile JN, Corson MA, Probstfield JL, Katz L, Peterson KA, Friedewald WT, Buse JB, Bigger JT, Gerstein HC, Ismail-Beigi F (2010a) Effects of intensive blood-pressure control in type 2 diabetes mellitus. N Engl J Med 362(17):1575–1585

ACCORD Study Group, Ginsberg HN, Elam MB, Lovato LC, Crouse JR 3rd, Leiter LA, Linz P, Friedewald WT, Buse JB, Gerstein HC, Probstfield J, Grimm RH, Ismail-Beigi F, Bigger JT, Goff DC Jr, Cushman WC, Simons-Morton DG, Byington RP (2010b) Effects of combination lipid therapy in type 2 diabetes mellitus. N Engl J Med 362(17):1563–1574

Action to Control Cardiovascular Risk in Diabetes Study Group, Gerstein HC, Miller ME, Byington RP, Goff DC Jr, Bigger JT, Buse JB, Cushman WC, Genuth S, Ismail-Beigi F, Grimm RH Jr, Probstfield JL, Simons-Morton DG, Friedewald WT (2008) Effects of intensive glucose lowering in type 2 diabetes. N Engl J Med 358(24):2545–2559

ADVANCE Collaborative Group, Patel A, MacMahon S, Chalmers J, Neal B, Billot L, Woodward M, Marre M, Cooper M, Glasziou P, Grobbee D, Hamet P, Harrap S, Heller S, Liu L, Mancia G, Mogensen CE, Pan C, Poulter N, Rodgers A, Williams B, Bompoint S, de Galan BE, Joshi R, Travert F (2008) Intensive blood glucose control and vascular outcomes in patients with type 2 diabetes. N Engl J Med 358(24):2560–2572

Alpert MA (2001) Obesity cardiomyopathy: pathophysiology and evolution of the clinical syndrome. Am J Med Sci 321(4):225–236

Ballard C, Schaffer S (1996) Stimulation of the Na+/Ca2+ exchanger by phenylephrine, angiotensin II and endothelin 1. J Mol Cell Cardiol 28(1):11–17

Barouch LA, Berkowitz DE, Harrison RW, O'Donnell CP, Hare JM (2003) Disruption of leptin signaling contributes to cardiac hypertrophy independently of body weight in mice. Circulation 108(6):754–759

Basta G, Schmidt AM, De Caterina R (2004) Advanced glycation end products and vascular inflammation: implications for accelerated atherosclerosis in diabetes. Cardiovasc Res 63(4):582–592

Belke DD, Betuing S, Tuttle MJ, Graveleau C, Young ME, Pham M, Zhang D, Cooksey RC, McClain DA, Litwin SE, Taegtmeyer H, Severson D, Kahn CR, Abel ED (2002) Insulin signaling coordinately regulates cardiac size, metabolism, and contractile protein isoform expression. J Clin Invest 109(5):629–639

Belke DD, Swanson EA, Dillmann WH (2004) Decreased sarcoplasmic reticulum activity and contractility in diabetic db/db mouse heart. Diabetes 53(12):3201–3208

Berentzen T, Sorensen TI (2006) Effects of intended weight loss on morbidity and mortality: possible explanations of controversial results. Nutr Rev 64(11):502–507

Berghöfer A, Pischon T, Reinhold T, Apovian CM, Sharma AM, Willich SN (2008) Obesity prevalence from a European perspective: a systematic review. BMC Public Health 8:200

Berrington de Gonzalez A, Hartge P, Cerhan JR, Flint AJ, Hannan L, MacInnis RJ, Moore SC, Tobias GS, Anton-Culver H, Freeman LB, Beeson WL, Clipp SL, English DR, Folsom AR, Freedman DM, Giles G, Hakansson N, Henderson KD, Hoffman-Bolton J, Hoppin JA, Koenig KL, Lee IM, Linet MS, Park Y, Pocobelli G, Schatzkin A, Sesso HD, Weiderpass E, Willcox BJ, Wolk A, Zeleniuch-Jacquotte A, Willett WC, Thun MJ (2010) Body-mass index and mortality among 1.46 million white adults. N Engl J Med 363(23):2211–2219

Bonds DE, Miller ME, Dudl J, Feinglos M, Ismail-Beigi F, Malozowski S, Seaquist E, Simmons DL, Sood A (2012) Severe hypoglycemia symptoms, antecedent behaviors, immediate consequences and association with glycemia medication usage: secondary analysis of the ACCORD clinical trial data. BMC Endocr Disord 12(1):5

Boudina S, Sena S, O'Neill BT, Tathireddy P, Young ME, Abel ED (2005) Reduced mitochondrial oxidative capacity and increased mitochondrial uncoupling impair myocardial energetics in obesity. Circulation 112(17):2686–2695

Boudina S, Bugger H, Sena S, O'Neill BT, Zaha VG, Ilkun O, Wright JJ, Mazumder PK, Palfreyman E, Tidwell TJ, Theobald H, Khalimonchuk O, Wayment B, Sheng X, Rodnick KJ, Centini R, Chen D, Litwin SE, Weimer BE, Abel ED (2009) Contribution of impaired myocardial insulin signaling to mitochondrial dysfunction and oxidative stress in the heart. Circulation 119(9):1272–1283

Brinkley TE, Hsu FC, Carr JJ, Hundley WG, Bluemke DA, Polak JF, Ding J (2011) Pericardial fat is associated with carotid stiffness in the Multi-Ethnic Study of Atherosclerosis. Nutr Metab Cardiovasc Dis 21(5):332–338

Calle EE, Thun MJ, Petrelli JM, Rodriguez C, Heath CW Jr (1999) Body-mass index and mortality in a prospective cohort of U.S. adults. N Engl J Med 341(15):1097–1105

Caterson ID, Finer N, Coutinho W, Van Gaal LF, Maggioni AP, Torp-Pedersen C, Sharma AM, Legler UF, Shepherd GM, Rode RA, Perdok RJ, Renz CL, James WP, SCOUT Investigators (2012) Maintained intentional weight loss reduces cardiovascular outcomes: results from the Sibutramine Cardiovascular OUTcomes (SCOUT) trial. Diabetes Obes Metab 14(6):523–530

Choi KM, Zhong Y, Hoit BD, Grupp IL, Hahn H, Dilly KW, Guatimosim S, Lederer WJ, Matlib MA (2002) Defective intracellular Ca(2+) signaling contributes to cardiomyopathy in Type 1 diabetic rats. Am J Physiol Heart Circ Physiol 283(4):H1398–H1408

Chugh PK, Sharma S (2012) Recent advances in the pathophysiology and pharmacological treatment of obesity. J Clin Pharm Ther 37(5):525–535

Cinti S, Mitchell G, Barbatelli G, Murano I, Ceresi E, Faloia E, Wang S, Fortier M, Greenberg AS, Obin MS (2005) Adipocyte death defines macrophage localization and function in adipose tissue of obese mice and humans. J Lipid Res 46(11):2347–2355

Ding EL, Malik VS (2008) Convergence of obesity and high glycemic diet on compounding diabetes and cardiovascular risks in modernizing China: an emerging public health dilemma. Global Health 4:4

Dokken BB (2008) The pathophysiology of cardiovascular disease and diabetes: beyond blood pressure and lipids. Diabetes Spectr 21(3):160–165

Duckworth W, Abraira C, Moritz T, Reda D, Emanuele N, Reaven PD, Zieve FJ, Marks J, Davis SN, Hayward R, Warren SR, Goldman S, McCarren M, Vitek ME, Henderson WG, Huang GD, VADT Investigators (2009) Glucose control and vascular complications in veterans with type 2 diabetes. N Engl J Med 360(2):129–139

Enriori PJ, Evans AE, Sinnayah P, Cowley MA (2006) Leptin resistance and obesity. Obesity (Silver Spring) 14(Suppl 5):254S–258S

Ferron L, Capuano V, Ruchon Y, Deroubaix E, Coulombe A, Renaud JF (2003) Angiotensin II signaling pathways mediate expression of cardiac T-type calcium channels. Circ Res 93(12):1241–1248

Fiordaliso F, Li B, Latini R, Sonnenblick EH, Anversa P, Leri A, Kajstura J (2000) Myocyte death in streptozotocin-induced diabetes in rats in angiotensin II-dependent. Lab Invest 80(4):513–527

Gudz TI, Tserng KY, Hoppel CL (1997) Direct inhibition of mitochondrial respiratory chain complex III by cell-permeable ceramide. J Biol Chem 272(39):24154–24158

Gunasegaram S, Haworth RS, Hearse DJ, Avkiran M (1999) Regulation of sarcolemmal Na(+)/H(+) exchanger activity by angiotensin II in adult rat ventricular myocytes: opposing actions via AT(1) versus AT(2) receptors. Circ Res 85(10):919–930

Guzzardi MA, Iozzo P (2011) Fatty heart, cardiac damage, and inflammation. Rev Diabet Stud 8(3):403–417

Herrero P, Peterson LR, McGill JB, Matthew S, Lesniak D, Dence C, Gropler RJ (2006) Increased myocardial fatty acid metabolism in patients with type 1 diabetes mellitus. J Am Coll Cardiol 47(3):598–604

Hossain P, Kawar B, El Nahas M (2007) Obesity and diabetes in the developing world – a growing challenge. N Engl J Med 356(3):213–215

Hue L, Taegtmeyer H (2009) The Randle cycle revisited: a new head for an old hat. Am J Physiol Endocrinol Metab 297(3):E578–E591

Iacobellis G, Bianco AC (2011) Epicardial adipose tissue: emerging physiological, pathophysiological and clinical features. Trends Endocrinol Metab 22(11):450–457

Iacobellis G, Ribaudo MC, Assael F, Vecci E, Tiberti C, Zappaterreno A, Di Mario U, Leonetti F (2003) Echocardiographic epicardial adipose tissue is related to anthropometric and clinical parameters of metabolic syndrome: a new indicator of cardiovascular risk. J Clin Endocrinol Metab 88(11):5163–5168

Iozzo P (2009) Viewpoints on the way to the consensus session: where does insulin resistance start? The adipose tissue. Diabetes Care 32(Suppl 2):S168–S173

Iozzo P (2011) Myocardial, perivascular, and epicardial fat. Diabetes Care 34(Suppl 2):S371–S379

Iozzo P, Chareonthaitawee P, Dutka D, Betteridge DJ, Ferrannini E, Camici PG (2002) Independent association of type 2 diabetes and coronary artery disease with myocardial insulin resistance. Diabetes 51(10):3020–3024

Kanda H, Tateya S, Tamori Y, Kotani K, Hiasa K, Kitazawa R, Kitazawa S, Miyachi H, Maeda S, Egashira K, Kasuga M (2006) MCP-1 contributes to macrophage infiltration into adipose tissue, insulin resistance, and hepatic steatosis in obesity. J Clin Invest 116(6):1494–1505

Kankaanpää M, Lehto HR, Pärkkä JP, Komu M, Viljanen A, Ferrannini E, Knuuti J, Nuutila P, Parkkola R, Iozzo P (2006) Myocardial triglyceride content and epicardial fat mass in human obesity: relationship to left ventricular function and serum free fatty acid levels. J Clin Endocrinol Metab 91(11):4689–4695

Kilhovd BK, Berg TJ, Birkeland KI, Thorsby P, Hanssen KF (1999) Serum levels of advanced glycation end products are increased in patients with type 2 diabetes and coronary heart disease. Diabetes Care 22(9):1543–1548

Lavie CJ, Amodeo C, Ventura HO, Messerli FH (1987) Left atrial abnormalities indicating diastolic ventricular dysfunction in cardiopathy of obesity. Chest 92(6):1042–1046

Lavie CJ, Milani RV, Ventura HO (2009) Obesity and cardiovascular disease: risk factor, paradox, and impact of weight loss. J Am Coll Cardiol 53(21):1925–1932

Lawlor DA, Davey Smith G, Ebrahim S, Thompson C, Sattar N (2005) Plasma adiponectin levels are associated with insulin resistance, but do not predict future risk of coronary heart disease in women. J Clin Endocrinol Metab 90(10):5677–5683

Lenzen M, Ryden L, Ohrvik J, Bartnik M, Malmberg K, Scholte Op Reimer W, Simoons ML, Euro Heart Survey Investigators (2006) Diabetes known or newly detected, but not impaired glucose regulation, has a negative influence on 1-year outcome in patients with coronary artery disease: a report from the Euro Heart Survey on diabetes and the heart. Eur Heart J 27(24):2969–2974

Libby P, Maroko PR, Braunwald E (1975) The effect of hypoglycemia on myocardial ischemic injury during acute experimental coronary artery occlusion. Circulation 51(4):621–626

Lindsay RS, Resnick HE, Zhu J, Tun ML, Howard BV, Zhang Y, Yeh J, Best LG (2005) Adiponectin and coronary heart disease: the Strong Heart Study. Arterioscler Thromb Vasc Biol 25(3):e15–e16

Matsuda M, Shimomura I, Sata M, Arita Y, Nishida M, Maeda N, Kumada M, Okamoto Y, Nagaretani H, Nishizawa H, Kishida K, Komuro R, Ouchi N, Kihara S, Nagai R, Funahashi T, Matsuzawa Y (2002) Role of adiponectin in preventing vascular stenosis. The missing link of adipo-vascular axis. J Biol Chem 277(40):37487–37491

Mazurek T, Zhang L, Zalewski A, Mannion JD, Diehl JT, Arafat H, Sarov-Blat L, O'Brien S, Keiper EA, Johnson AG, Martin J, Goldstein BJ, Shi Y (2003) Human epicardial adipose tissue is a source of inflammatory mediators. Circulation 108(20):2460–2466

McGavock JM, Lingvay I, Zib I, Tillery T, Salas N, Unger R, Levine BD, Raskin P, Victor RG, Szczepaniak LS (2007) Cardiac steatosis in diabetes mellitus: a 1H-magnetic resonance spectroscopy study. Circulation 116(10):1170–1175

Mercer BN, Morais S, Cubbon RM, Kearney MT (2012) Diabetes mellitus and the heart. Int J Clin Pract 66(7):640–647

Messerli FH, Nunez BD, Ventura HO, Snyder DW (1987) Overweight and sudden death. Increased ventricular ectopy in cardiopathy of obesity. Arch Intern Med 147(10):1725–1728

Nobe S, Aomine M, Arita M, Ito S, Takaki R (1990) Chronic diabetes mellitus prolongs action potential duration of rat ventricular muscles: circumstantial evidence for impaired Ca2+ channel. Cardiovasc Res 24(5):381–389

Ogden CL, Yanovski SZ, Carroll MD, Flegal KM (2007) The epidemiology of obesity. Gastroenterology 132(6):2087–2102

Ouedraogo R, Gong Y, Berzins B, Wu X, Mahadev K, Hough K, Chan L, Goldstein BJ, Scalia R (2007) Adiponectin deficiency increases leukocyte-endothelium interactions via upregulation of endothelial cell adhesion molecules in vivo. J Clin Invest 117(6):1718–1726

Pereira L, Matthes J, Schuster I, Valdivia HH, Herzig S, Richard S, Gómez AM (2006) Mechanisms of [Ca2+]i transient decrease in cardiomyopathy of db/db type 2 diabetic mice. Diabetes 55(3):608–615

Peterson LR, Herrero P, Schechtman KB, Racette SB, Waggoner AD, Kisrieva-Ware Z, Dence C, Klein S, Marsala J, Meyer T, Gropler RJ (2004) Effect of obesity and insulin resistance on myocardial substrate metabolism and efficiency in young women. Circulation 109(18):2191–2196

Pischon T, Girman CJ, Hotamisligil GS, Rifai N, Hu FB, Rimm EB (2004) Plasma adiponectin levels and risk of myocardial infarction in men. JAMA 291(14):1730–1737

Privratsky JR, Wold LE, Sowers JR, Quinn MT, Ren J (2003) AT1 blockade prevents glucose-induced cardiac dysfunction in ventricular myocytes: role of the AT1 receptor and NADPH oxidase. Hypertension 42(2):206–212

Raha S, Robinson BH (2000) Mitochondria, oxygen free radicals, disease and ageing. Trends Biochem Sci 25(10):502–508

Romero-Corral A, Sierra-Johnson J, Lopez-Jimenez F, Thomas RJ, Singh P, Hoffmann M, Okcay A, Korinek J, Wolk R, Somers VK (2008) Relationships between leptin and C-reactive protein with cardiovascular disease in the adult general population. Nat Clin Pract Cardiovasc Med 5(7):418–425

Schenk S, Saberi M, Olefsky JM (2008) Insulin sensitivity: modulation by nutrients and inflammation. J Clin Invest 118(9):2992–3002

Schenkeveld L, Magro M, Oemrawsingh RM, Lenzen M, de Jaegere P, van Geuns RJ, Serruys PW, van Domburg RT (2012) The influence of optimal medical treatment on the 'obesity paradox', body mass index and long-term mortality in patients treated with percutaneous coronary intervention: a prospective cohort study. BMJ Open 2:e000535

Scholte AJ, Schuijf JD, Kharagjitsingh AV, Jukema JW, Pundziute G, van der Wall EE, Bax JJ (2008) Prevalence of coronary artery disease and plaque morphology assessed by multi-slice computed tomography coronary angiography and calcium scoring in asymptomatic patients with type 2 diabetes. Heart 94(3):290–295

Schramm TK, Gislason GH, Køber L, Rasmussen S, Rasmussen JN, Abildstrøm SZ, Hansen ML, Folke F, Buch P, Madsen M, Vaag A, Torp-Pedersen C (2008) Diabetes patients requiring glucose-lowering therapy and nondiabetics with a prior myocardial infarction carry the same cardiovascular risk: a population study of 3.3 million people. Circulation 117(15):1945–1954

Schwartz MW, Seeley RJ, Campfield LA, Burn P, Baskin DG (1996) Identification of targets of leptin action in rat hypothalamus. J Clin Invest 98(5):1101–1106

Shimabukuro M (2009) Cardiac adiposity and global cardiometabolic risk: new concept and clinical implication. Circ J 73(1):27–34

Stratton IM, Adler AI, Neil HA, Matthews DR, Manley SE, Cull CA, Hadden D, Turner RC, Holman RR (2000) Association of glycaemia with macrovascular and microvascular complications of type 2 diabetes (UKPDS 35): prospective observational study. BMJ 321(7258):405–412

Szmitko PE, Teoh H, Stewart DJ, Verma S (2007) Adiponectin and cardiovascular disease: state of the art? Am J Physiol Heart Circ Physiol 292(4):H1655–H1663

Turko IV, Murad F (2003) Quantitative protein profiling in heart mitochondria from diabetic rats. J Biol Chem 278(37):35844–35849

UKPDS (1998) Intensive blood-glucose control with sulphonylureas or insulin compared with conventional treatment and risk of complications in patients with type 2 diabetes (UKPDS 33). UK Prospective Diabetes Study (UKPDS) Group. Lancet 352(9131):837–853

van Marken Lichtenbelt WD, Vanhommerig JW, Smulders NM, Drossaerts JM, Kemerink GJ, Bouvy ND, Schrauwen P, Teule GJ (2009) Cold-activated brown adipose tissue in healthy men. N Engl J Med 360(15):1500–1508

Wild S, Roglic G, Green A, Sicree R, King H (2004) Global prevalence of diabetes: estimates for the year 2000 and projections for 2030. Diabetes Care 27(5):1047–1053

Wu AH, Pitt B, Anker SD, Vincent J, Mujib M, Ahmed A (2010) Association of obesity and survival in systolic heart failure after acute myocardial infarction: potential confounding by age. Eur J Heart Fail 12(6):566–573

Yaari S, Goldbourt U (1998) Voluntary and involuntary weight loss: associations with long term mortality in 9,228 middle-aged and elderly men. Am J Epidemiol 148(6):546–555

Young ME, Laws FA, Goodwin GW, Taegtmeyer H (2001) Reactivation of peroxisome proliferator-activated receptor alpha is associated with contractile dysfunction in hypertrophied rat heart. J Biol Chem 276(48):44390–44395

Yu Q, Gao F, Ma XL (2010) Insulin says NO to cardiovascular disease. Cardiovasc Res 89(3):516–524

Zhang X, Chen C (2012) A new insight of mechanisms, diagnosis and treatment of diabetic cardiomyopathy. Endocrine 41(3):398–409

Hyper- and Dyslipoproteinemias

Karam M. Kostner and Gert M. Kostner

Abstract

Hyper- and dyslipoproteinemias are hallmarks for atherosclerosis, cardiovascular diseases, and stroke. Under normal physiological conditions, lipids are transported by three major lipoprotein classes: triglyceride-rich (chylomicrons, VLDL), cholesterol-rich (LDL, Lp(a)), and protein-rich (HDL) lipoproteins. Dietary fat after absorption is incorporated into chylomicrons that are catabolized in the "exogenous pathway." The endogenous pathway of lipoproteins starts with the secretion of VLDL from liver, the hydrolysis by lipoprotein lipase, and the formation of the final product, LDL. LDL are thought to deliver cholesterol to peripheral organs via the LDL receptor. A low LDL-receptor activity is characteristic for high plasma LDL, and vice versa. Due to their relatively long residence time in circulation, LDL are modified by oxidative stress and high glucose levels. These particles cannot bind to the LDL receptor and must be cleared from circulation by "scavenger receptors" of macrophages. An overload of such pathogenic lipoproteins gives rise to foam cell formation, atherosclerotic plaques, and occlusion of arteries. HDL on the other hand are the good guys as they take up cholesterol from periphery and transport them to the liver (reverse cholesterol transport). The liver converts this fraction of cholesterol into bile acids that are secreted together with free cholesterol into bile. There exist additional very atherogenic lipoproteins such as Lp(a), small dense LDL, or pre-β LDL. A derangement in the well-balanced lipoprotein metabolism leads to hyper- and dyslipoproteinemias, metabolic syndrome, and atherosclerosis.

K.M. Kostner
Department of Cardiology,
Mater Adult Hospital, University of Queensland, Brisbane, QLD, Australia

G.M. Kostner (✉)
Department of Molecular Biology and Biochemistry,
Medical University of Graz, Harrachgasse 21/III, 8010 Graz, Austria
e-mail: gerhard.kostner@medunigraz.at

Keywords

Atherosclerosis • Lipoproteins • Metabolism • Cholesterol balance • Lp(a) • Enzymes • Lipoprotein receptors

3.1 Plasma Lipoproteins: Structure and Composition

The main plasma lipids, including triglycerides (TG), free cholesterol (FC), cholesteryl ester (CE), and partly also phospholipids (PL), are water insoluble and thus are transported in blood in the form of lipoproteins (Lp) (reviewed by Kostner 1983). The shell of Lp consists of a characteristic protein moiety in addition to the PL lecithin and sphingomyelin and of FC (Fig. 3.1). Several classification systems for Lp that may be applied simultaneously as appropriate exist (Courtney and Janssen 2006; Myers et al. 1984; Brunzell et al. 1978). Table 3.1 summarizes the most relevant ones for this chapter. Thus, lipoproteins are denominated according to their composition as TG-rich Lp comprising chylomicrons, very-low-density lipoproteins (VLDL), and intermediate-density lipoproteins (IDL). To the class of CE-rich

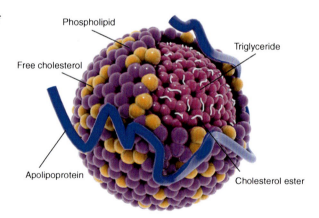

Fig. 3.1 Schematic view of a plasma lipoprotein

Table 3.1 Lipoprotein classes of human plasma

Lp class	Synonym	Major apo-Lp	Atherogenicity
I. Triglyceride-rich lipoproteins			
Chylomicrons	–	B-48, E, CI, CII, CIII	–
Chylomicron remnants	–	B-48, E	+++
VLDL	Pre-β-Lp	B-100, E, CI, CII, CIII	+
IDL	Floating β-Lp	B-100, E	+++
II. Cholesterol- and cholesteryl ester-rich lipoproteins			
LDL	β-Lp	B-100	++
β-VLDL	Dys-β-Lp	B-100, E	+++
Lp(a)	LPA	B-100, apo(a)	+++
III. ApoA-containing lipoproteins			
HDL2	–	AI, AII, C	--
HDL3	–	AI, AII, C	--
Pre-β-HDL	γ-HDL	AI	---

3 Hyper- and Dyslipoproteinemias

Table 3.2 Apolipoproteins of human plasma

Apo-Lp	Synonym	Function
AI	–	LCAT activation, reverse cholesterol transport
AII	–	Unknown
AIV		TG absorption
AV		TG catabolism
B-100	–	VLDL synthesis, ligand for the LDL-R
B-48		Chylomicron synthesis
CI		
CII		Activation of LPL
CIII		Inhibition of LPL
D	AIII	
E		Cholesterol efflux, ligand for the LDL-R and LRP
H	β2-GP-I	TG metabolism, binding of negatively charged PL
M		Transport of sphingosine-1-P

Abbreviations: *LCAT* lecithin-cholesterol acyltransferase, *LRP* LDL-receptor-related protein

lipoproteins belong low-density lipoproteins (LDL), lipoprotein(a) (Lp(a)), and β-VLDL, a fraction that is elevated in dys-β-lipoproteinemic patients. High-density lipoproteins (HDL) finally are rich in proteins and phospholipids (PL). LDL are also called β-Lp, VLDL are called pre-β-Lp, and HDL are called alpha-Lp. According to their major protein content, apoA-containing Lp are HDL and their subclasses −2 and −3; apoB-containing Lp are chylomicrons, VDL, IDL, LDL, and Lp(a); and apoC-containing Lp are mainly chylomicrons, VLDL, and HDL (Alaupovic et al. 1985). There exist in addition numerous other minor lipoprotein classes that will not be addressed in this chapter and are of little clinical relevance. One exception is the so-called pre-β-HDL, a lipoprotein that is believed to play a major role in the reverse cholesterol transport (RCT). Another Lp that is important for our considerations is the so-called small dense (sd) LDL, a fraction that is mainly found in hypertriglyceridemic plasma with a density close to HDL1.

The characteristic features of the different Lp classes are the various apo-Lp despite the lipid part (Dominiczak and Caslake 2011). Apo-Lp display very defined functions: they stabilize the solution properties of Lp (apoA and apoB), they are activators or inhibitors of enzymes (apoAI, apoCII, and CIII), they are ligands for binding to Lp receptors (apoB, apoE), and they are indispensable for the biosynthesis and secretion of lipids (apoB, apoE). The major apo-Lp and their functions are summarized in Table 3.2.

3.2 Methods for the Preparation and Analysis of Lipoproteins

Excellent reviews of methods that are currently applied for Lp separation and analysis can be found in the literature (Tadey and Purdy 1995; Warnick 1986, 1990), and thus, we will give here only a brief summary. The method of choice for preparing lipoprotein density fractions is the ultracentrifuge using swinging-bucket, fixed-angle, or vertical rotors. There are several 100,000×g forces necessary to separate Lp classes. This is achieved either in a density gradient or by stepwise increasing the background density to 1.006 (for chylomicrons and VLDL), 1.063 (for LDL), 1.125 (for Lp(a) and HDL2), and 1.21 (for HDL3). These densities are adjusted by adding solid NaCl or NaBr to the plasma. In continuous density gradients, VLDL, IDL, LDL, and HDL may be separated into five or even more subfractions each. For further purification, crude density fractions are chromatographed over preparative columns with size exclusion material (e.g., Sephadex or Bio-Gel) or ion exchangers, e.g., DEAE cellulose. An elegant method for purifying Lp, in particular if the metabolism is studied, is affinity columns using immune-specific adsorbers, lectins, or other resins. In addition, crude Lp fractions may be obtained by selective precipitation methods using ammonium sulfate, phosphotungstic acid, heparin, or dextran sulfate (Bachorik and Albers 1986).

A complete chemical analysis of lipoproteins may be performed after purification by the abovementioned methods. Since these are time consuming and not always necessary, Lp are analytically separated by different electrophoretic methods (cellulose acetate, agarose, polyacrylamide gels) following staining and densitometric quantification.

The most widely applied methods, however, are precipitation of apoB-containing Lp by heparin, dextran sulfate, or phosphotungstic acid in the presence of bivalent cations (Mg^{2+}, Ca^{2+}, or Mn^{2+}) followed by the enzymatic analysis of cholesterol in the precipitate or in the supernatant (Rifai et al. 1992).

The chemical analysis of Lp is mostly performed by the Lowry or Bartelet protein assay and the enzymatic quantification of individual lipids. A thorough analysis in specialized laboratories succeeds by proteomics and lipidomics using gas chromatography–mass spectrometry (Gordon et al. 2010; Hoofnagle and Heinecke 2009; Kontush and Chapman 2010; Sun et al. 2010). There is on the other hand an elegant NMR-based analytical method that separates the plasma lipoprotein spectrum into numerous subfractions mostly based on particle size. This method is commercialized and further information might be found at http://www.lipofit.de/index_engl.html.

3.3 Metabolism of Lipoproteins

The metabolism of plasma lipoproteins has been reviewed in numerous publications (Boren et al. 2012; Choi and Ginsberg 2011; Dolphin 1985; Kolovou et al. 2011; Nakajima et al. 2011; Pownall et al. 1979; Xiao et al. 2011). There are numerous enzymes and other functional proteins involved in Lp metabolism.

3.3.1 Enzymes and Exchange/Transfer Proteins

Lipoprotein lipase (*LPL*) hydrolyzes Lp-bound triglycerides, liberating free fatty acids (FFA), monoglycerides, and glycerol. LPL cleaves TG in chylomicrons and

VLDL and needs apoCII as a cofactor. ApoCIII on the other hand inhibits LPL activity. LPL is synthesized in muscle cells and adipose tissue and transported by specific proteins to the surface of endothelial cells where they are non-covalently bound to proteoglycans (mostly heparan sulfate). Upon intravenous injection of heparin, LPL (in addition to the other lipases mentioned below) is liberated and may be assayed in plasma in the form of PHLA (postheparin lipolytic activity). Under normal conditions, the majority of LPL and hepatic lipase (HL), but only approx. 60 % of endothelial lipase (EDL), is bound to cell surfaces. Mutated forms of these lipases have been found that show a reduced proteoglycan binding and cause derangements in lipoprotein metabolism.

Hepatic Lipase and Endothelial Lipase: HL is mainly expressed in the liver and shows a similar metabolism as LPL; however, HL has a much greater affinity to IDL as well as HDL and plays a major role in the generation of LDL. The substrate of endothelial lipase (EDL) is mainly phosphatidylcholine (lecithin) of HDL, and overexpression of EDL is associated with reduced plasma HDL concentrations.

Lecithin-Cholesterol Acyltransferase (LCAT): LCAT acts mainly on HDL and is responsible for the formation of the major amount of CE in plasma. It transfers the fatty acid in position-2 of lecithin to FC generating CE and lysolecithin. LCAT is activated by apoAI and probably also by other apo-Lp. For example, also a so-called β-LCAT acting on LDL has been found, yet the exact nature of β-LCAT has not been defined so far.

Cholesteryl Ester Exchange/Transfer Protein (CETP): CETP exchanges core lipids (CE and TG) between different Lp fractions. Its major function is the transfer of CE synthesized by LCAT in HDL to LDL or IDL and the exchange of CE for TG. TG transferred to HDL in this pathway are hydrolyzed, giving rise to smaller, CE-poor HDL fractions. Inhibition of CETP is associated with strikingly elevated HDL-C concentrations.

Phospholipid Exchange/Transfer Protein (PLTP): PLTP transfers surface components of Lp, mainly lecithin and sphingomyelin, between different Lp classes and causes a redistribution of Lp. For example, if HDL3 is incubated with PLTP, HDL2 and very-high-density lipoproteins are generated. In addition PLTP has been found to transfer alpha-tocopherol (Vit. E) from one Lp class to another or from lipoproteins to cell membranes (Kostner et al. 1995) (Table 3.3).

Table 3.3 Enzymes and transfer proteins in Lp metabolism

Protein	Mode of action	Substrate	Major organ
LPL	TG hydrolysis	VLDL, chylomicrons	Adipose tissue and muscle
HL	TG hydrolysis	IDL, HDL	Liver
EDL	PL-hydrolysis	HDL	Endothelial cells
LCAT	Transfer of FA from lecithin to FC	HDL	Liver
CETP	Transfer of CE to VLDL and LDL, exchange of TG with CE	HDL	Liver
PLTP	Exchange and transfer of PL and Vit. E		Liver

Table 3.4 Lipoprotein receptors

Receptor	Organs	Function
LDL-R	All organs, major organ: liver	Catabolism of LDL
LRP	Liver	Catabolism of remnants
VLDL-R	Adipose tissue	Catabolism of TG-rich Lp
SR-A	Macrophages	Removal of atherogenic Lp
SR-BI	Macrophages, liver	Removal of atherogenic Lp, reverse cholesterol transport
LOX-1	Macrophages	Removal of oxidized Lp

3.3.2 Lipoprotein Receptors

Numerous specific lipoprotein receptors have been characterized that play important roles in Lp metabolism (Havel 1995; Herz and Willnow 1995; Linton and Fazio 1999; Schneider et al. 1997) (Table 3.4). The best known and probably most important one in Lp metabolism is the LDL receptor first described by Brown and Goldstein (2009). LDL receptors are present on the surface of almost any cell and have a high affinity to apoB- and apoE-containing lipoproteins. Equally important are the remnant receptors and LDL-R-related protein (LRP) found mostly on liver cells. Another receptor is a VLDL-specific receptor in adipose tissue and muscle, in addition to several not fully characterized HDL receptors (see Sect. 3.3.5). Finally, there are several so-called scavenger receptors (SR) known to bind mostly oxidized and modified or negatively charged lipoproteins. The most important ones are SR-A, SR-BI (Yamada et al. 1998), and CD-36. CD-36 is also known as the free fatty acid (FFA) receptor. Another scavenger receptor that is important for the removal of oxidized Lp is the lectin-like oxidized LDL receptor-1 (LOX-1), reviewed by Lu et al. (2011). All the scavenger receptors are found mostly on macrophages but play also a role in the intestine, the liver, and adipose tissue.

3.3.3 The Exogenous Pathway of Lipoprotein Metabolism

This pathway is summarized in Fig. 3.2. Dietary lipids and lipid-soluble vitamins are partly hydrolyzed by pancreatic lipases and esterases in the intestine. Under normal healthy conditions the majority of dietary TG are absorbed after hydrolysis, whereas only 30–60 % of the ingested cholesterol is taken up and the rest is secreted into feces. This latter pathway is under genetic control. For the absorption of lipids, bile acids are essential since they emulsify fat and make it accessible to pancreatic lipases (Xiao et al. 2011). FFA and FC generated in this way are utilized in the enterocytes for TG and CE biosynthesis and chylomicron formation. The prominent protein in chylomicrons is apoB-48 that has 48 % of the size of apoB-100, the major protein in LDL. Chylomicrons enter the chyle and via ductus thoracicus they reach the blood stream where they are immediately hydrolyzed by LPL, giving rise to FFA formation. FFA bind to serum albumin and are taken up by skeletal and cardiac muscle, adipose tissue, and liver in a CD-36-mediated process and further

3 Hyper- and Dyslipoproteinemias

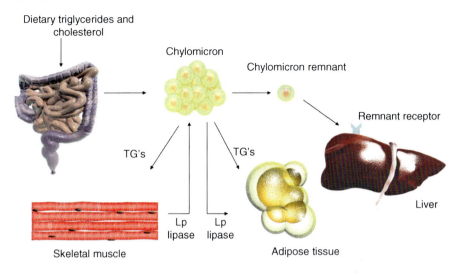

Fig. 3.2 Exogenous pathway of lipid metabolism

metabolized intracellularly. The core remnants of chylomicrons are taken up by specific receptors on liver cells, whereas surface remnants enter the HDL cycle.

3.3.4 The Endogenous Pathway of Lipoprotein Metabolism

The central organ for this pathway is the liver (Olofsson et al. 2007). FFA entering liver cells are utilized for TG biosynthesis. The liver also takes up cholesterol via RTC and also synthesizes TG, FC, CE, and PL de novo. These lipids are packed in a rather complicated sequence of events into large VLDL that are secreted from liver into circulation. For VLDL biosynthesis, microsomal TG transfer protein (MTP) is essential. VLDL have as major protein apoB-100 that is continuously biosynthesized. ApoB-100 that is not loaded with lipids undergoes proteolytic digestion (Olofsson et al. 1999). Thus, VLDL biosynthesis and secretion are not controlled via transcription of apoB but rather by numerous other metabolic variables that are not fully explored. VLDL circulating in blood undergo a similar pathway as chylomicrons, namely, hydrolysis of their TG into FFA by LPL, yet the half-life of VLDL is some 10–20 h, whereas that of chylomicrons is only minutes up to 1 h. In the sequence of VLDL-TG hydrolysis, LPL is responsible for IDL formation that in turn is converted by HL into LDL. LDL have a half-life of 2–3 days and are catabolized via LDL-R mainly by the liver but also by other organs. Because of the relatively long half-life of LDL, there are several processes that modify LDL by oxidation, glycation, and carbamylation or by binding of other compounds causing an excess of negative charges. This is the recognition signal for scavenger receptors present on macrophages and liver cells that are responsible for the catabolism of modified LDL (Fig. 3.3).

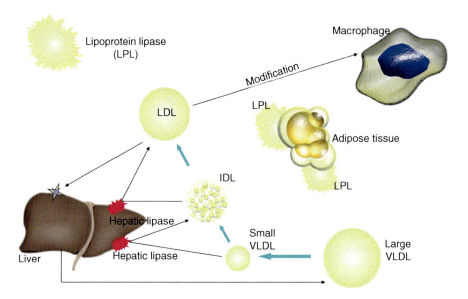

Fig. 3.3 Endogenous pathway of lipid metabolism

3.3.5 The Reverse Cholesterol Transport

For a long time, the role of HDL in Lp metabolism and atherogenesis was strongly underestimated. According to current theories, however, HDL play a central role in the removal of excess cholesterol mainly from peripheral tissue, explaining part of its role as "good cholesterol." As shown in Fig. 3.4, HDL takes up cholesterol from cell membranes of peripheral cells, notably macrophages and foam cells. There are several proteins involved in this pathway such as ABC transporters, NPC protein, SRB-I, and others. Since intracellular FC is cytotoxic, any surplus of cellular cholesterol is either stored in the form of CE or excreted ("effluxed") to apoAI-containing particles. The exact mechanism is not fully understood, yet we know that there are at least two ABC transporters involved in this process: ABCA-1 and ABCG-1 (Ye et al. 2011). ABCA-1 transfers FC to nascent apoAI-PL particles (pre-β-HDL), and thus, pre-β-HDL is considered to be the most anti-atherogenic subfraction in HDL. ABCG-1 on the other hand transfers FC to normal HDL and appears to play a less anti-atherogenic role. The expression of these ABC transporters is heavily stimulated by the nuclear receptor LXR. The FC accumulated in this pathway in HDL is converted to CE by LCAT and transferred to VLDL and LDL by CETP. LDL and VLDL remnants are taken up by receptor-mediated mechanisms mainly by the liver.

Liver cells on the other hand expose the scavenger receptor SR-BI that mediates the so-called selective uptake of HDL-CE (but also LDL-CE), leaving the CE-poor particles behind in the circulating blood. The cholesterol taken up from HDL into liver cells is preferentially converted into bile acids that are secreted into bile together with FC with the aid of specific transporter proteins.

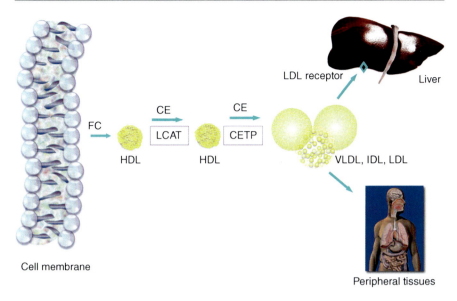

Fig. 3.4 Reverse cholesterol transport. *FC* free cholesterol, *TG* triglycerides, *CE* cholesterol esters, *LCAT* lecithin cholesterol acyl transferase, *CETP* cholesterol ester transfer protein

3.3.6 Total Body Cholesterol Balance

Cholesterol is important as cell membrane component in addition to its role as precursor of bile acids and steroid hormones. Although almost any human tissue may synthesize cholesterol de novo from acetyl-CoA by a process consisting of more than 35 enzyme-catalyzed steps, endocrine organs and liver activate specific pathways to save energy and take up Lp cholesterol from blood. In adults the dietary need of FC serving these pathways is approx. 300 mg/day, an amount that is present in the yolk of 1.5–2 eggs (Fig. 3.5).

There is an ongoing debate on the exact mechanism of cholesterol absorption. The current assumption is that FC emulsified in mixed micelles together with PL, FFA, and bile salts is taken up by enterocytes in an active transport involving several proteins, the most important being NPC-1L1 (Fig. 3.6). Cholesterol absorption inhibitors such as ezetimibe® interfere with this step of cholesterol absorption, thereby lowering plasma cholesterol. There are two additional ABC transporters involved in the subsequent step, ABC-G5 and ABC-G8 acting together (Fig. 3.6). This heterodimer transporter shovels FC out of the enterocytes back into the small intestinal lumen together with plant sterols that are only minimally absorbed under normal conditions. Genetic defects in ABC-G5 and/or ABC-G8 lead to sitosterolemia, a condition associated with extremely high plasma plant sterol concentrations and elevated plasma cholesterol. In any case, dietary plant sterols given in doses of 1–3 g/day inhibit cholesterol absorption in this way but are hardly absorbed. Whether or not plant sterols accumulating in blood and tissues are equally or even more atherogenic than FC is currently under heavy debate. As shown in Fig. 3.5,

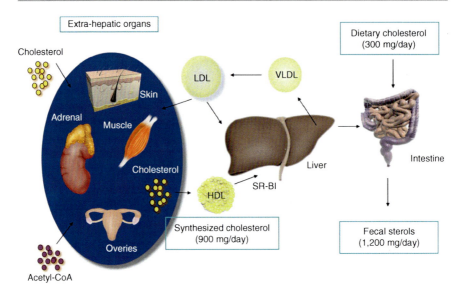

Fig. 3.5 Cholesterol balance

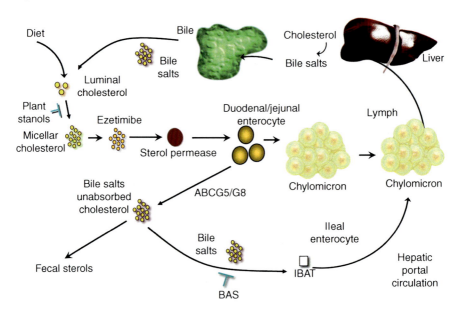

Fig. 3.6 Mechanism of intestinal lipid absorption

a 70 kg adult taking up 300 mg/day of dietary FC secretes 1,200 mg/day of neutral and acidic sterols into feces, indicating that under steady-state conditions, 900 mg of cholesterol is de novo biosynthesized. Parenchymal liver cells in fact contain also ABC-G5/ABC-G8 transporters that are responsible for the secretion of FC into bile. These FC in addition to bile salts are reabsorbed in the enterohepatic cycle. Interference with the reabsorption of FC by ezetimibe® or of bile salts by specific sequestrants in this cycle is one pharmacological strategy to lower plasma cholesterol. Recently, a new pathway of intestinal cholesterol excretion was postulated called TICE (trans-intestinal cholesterol excretion) involving ABCA-1 and HDL (Vrins 2010).

3.3.7 Cellular Cholesterol Balance

The cholesterol metabolism at cellular level in different tissues is individually and tightly regulated depending on whether the tissue synthesizes and secretes Lp, produces steroid hormones, scavenges cholesterol from atherosclerotic plaques, or utilizes FC for membrane stabilization. Common to most of the cells is the "Brown and Goldstein" pathway (Goldstein and Brown 2009). In the absence of extracellular cholesterol, cells synthesize FC from acetyl-CoA under the control of the key enzymes HMG-CoA reductase, HMG-CoA synthase, and sterol epoxidase. The expression of these enzymes is under the control of the transcription factor SREBP-2 that is regulated by LXR. The molecular mechanism of this multistep regulatory pathway has been clarified. In the presence of extracellular Lp cholesterol, receptor-mediated or selective cholesterol uptake takes place. Important in these processes are the LDL-R, SR-B1, and several enzymes. LDL is bound to the LDL-R and internalized, and CE is degraded in lysosomes, yielding FC that in turn regulates the activity of LDL-R, HMG-CoA reductase, and acyl-CoA cholesterol acyltransferase (ACAT). Cellular cholesterol might be stored as CE, secreted in the form of apoB- or apoE-containing lipoproteins or effluxed via ABC transporters. This latter pathway is highly relevant for the deposition or mobilization of cholesterol in atherosclerotic plaques.

3.4 Dys- and Hyper-/Hypolipoproteinemias (D&HLP)

D&HLP are hallmarks in coronary heart disease, myocardial infarction, and stroke (Hazzard et al. 1973). The term "dyslipoproteinemia" is commonly used if abnormal Lp not found in healthy fasting plasma accumulate in plasma. In hyper- or hypolipoproteinemia, normal plasma lipoproteins are elevated or reduced in concentrations. Since a complete description of these diseases by far exceeds the scope of this chapter, we will focus on the most common forms of D&HLP that are relevant for the present considerations.

For practical purposes, D&HLP are classified into hypercholesterolemia and familial defective apoB-100, hypertriglyceridemias, mixed hyperlipidemia including familial combined hyperlipidemia, hypo-HD-lipoproteinemia, and elevated Lp(a) (Table 3.5). D&HLP in addition are divided into primary (familial, caused by mutations and polymorphisms) and secondary diseases.

Table 3.5 Dys- and hyperlipoproteinemias

Common name	Synonym	Defect and etiology	Atherogenicity
I. Hypercholesterolemias			
Fam. hypercholesterolemia	Type IIA HLP	Mutation in LDL-R	+++
PCSK9 defect	Gain-of-function PCSK9 mutation	Increased degradation of LDL-R	+++
B-3500 variant	FDB	Mutation in ApoB-100	++
II. Hypertriglyceridemias			
LPL defect	Type I HLP	Genetic defect in LPL	–
ApoCII defect	Type I HLP	Absence of apoCII, the activator of LPL	–
Hyper-VLDL	Type IV	Overproduction of VLDL	++
III. Mixed hyperlipidemias			
Generalized HLP	Type V	Elevation of all apoB-containing lipoproteins	++
Mixed HLP	Type IIB	Elevation of VLDL and LDL, unknown mechanism	++
Fam. dys-β-lipoproteinemia	Type III	Mutations in apoE; most patients have APOE2	+++
Fam. combined HLP	Type IIB, III, IV, V	Family members have variable phenotypes	++
IV. Hypo-HDL			
Tangier diseases	An-alpha-lipoproteinemia	Genetic defect in ABCA-1	+
Hypo-alpha-lipoproteinemia		Metabolic defects involving ABCA-1, EDL, CETP, HL	+++
V. Hyper-Lp(a)			
Elevated Lp(a)		Genetically mediated overproduction of apo(a)	+++
Secondary hyper-Lp(a)		In patients with nephrotic syndrome and end-stage renal disease, Lp(a) is 3- to 5-fold increased	+++
		FH and hypothyroid patients have significantly elevated Lp(a)	

3.4.1 Hypercholesterolemia and Familial Defective ApoB-100

3.4.1.1 Primary Hypercholesterolemias

FH, Type IIA: These genetic diseases without doubt are most relevant for atherogenesis. In the classical form of familial hypercholesterolemia (FH, also called Type IIA according to Fredrickson), the LDL-R is mutated and either defective or functionally reduced. There are currently >1,500 mutations and polymorphisms known that affect the LDL-R activity. In this disease, liver produces an excess of apoB-containing Lp that are secreted and not properly reabsorbed into the liver. As a consequence, cholesterol biosynthesis is not normally downregulated in the Brown and Goldstein pathway, and in turn, LDL cholesterol accumulates in plasma. Heterozygous FH patients exhibit plasma cholesterol levels of 350 mg/dl and more (normal value 200 ± 20 mg/dl), and in homozygous FH patients up to 1,000 mg/dl cholesterol may be observed. Such patients are at an extremely high atherogenic risk. FH occurs with a frequency of approx. 1:500 in the western population.

Hobbs from the group of Brown and Goldstein was first to recognize another genetic defect leading to D&HLP, namely, mutations in PCSK9 (Horton et al. 2009). PCSK9 is a protease that cleaves and inactivates the LDL-R and reduces its recycling from lysosomes to cell membranes. There are "gain-of-function and loss-of-function" mutations of PCSK9 known. Mutations that cause a gain of function degrade LDL-R more efficiently, leading to a similar phenotype as compared to Type IIA FH caused by LDL-R mutations. PCSK9 is believed to be a promising target for pharmacological intervention in hypercholesterolemic states.

There are at least two additional mutations described in proteins involved in the LDL-R pathway giving rise to primary hypercholesterolemia.

Familial Defective β-Lipoproteinemia (FDB): Since apoB-100 is the ligand for the LDL-R, it is evident that mutations in apoB may also interfere with proper receptor binding and catabolism of LDL. In fact there is an apoB mutation with a rather high frequency, called ApoB-3500 mutation. In this mutation, the amino acid Arg at position 3500 is substituted by Glu, causing a reduced LDL binding to its receptor. Hypercholesterolemia under this condition is less pronounced since VLDL and their remnants possess in addition to apoB-100 also apoE that has a high affinity to LDL-R and to LRP that catabolizes these Lp.

To make the picture complete, there are several mutations described in the literature affecting the apoB-100 size. Due to mutations there are stop codons introduced at variable positions in the apoB gene, giving rise to truncated forms of apoB, called, for example, apoB-75, apoB-45, or apoB-28, and more. Interestingly, these mutations cause hypo-β-lipoproteinemias that protect from atherosclerosis.

In the very rare disease *a-beta-lipoproteinemia*, the apoB gene is completely normal, yet MTP is not functional, and in turn, no chylomicrons, VLDL, or LDL are found in plasma. These patients suffer mostly from deficiencies of fat-soluble vitamins that cannot be absorbed in the absence of apoB. MTP is currently a target for developing hypolipidemic drugs.

3.4.1.2 Secondary Hypercholesterolemia

As thyroid hormones play an important role in cholesterol biosynthesis and LDL-R-mediated catabolism of apoB-containing lipoproteins, hypothyroidism causes significantly elevated plasma cholesterol levels that may be treated by hormone substitutions.

3.4.2 Hypertriglyceridemias

3.4.2.1 Primary Hypertriglyceridemias

TG in plasma are hydrolyzed by LPL and HL. Thus, any abnormally low activity of these enzymes leads to TG accumulation in plasma and hypertriglyceridemias (HTG). The classical familial HTG, also called "Type I" according to the Fredrickson nomenclature (Fredrickson et al. 1967), is the absence of functional LPL. There are several known mutations causing this disease. Plasma TG concentrations in Type I HLP may reach 10,000 mg/dl and more. These patients suffer from pancreatitis at early age. Treatment of this disease succeeds with strict diet free of long-chain TG that are substituted by medium-chain TG, the latter being catabolized by the portal vein. Since LPL needs apoCII as a cofactor, mutations in apoCII have also been found to lead to similar phenotypes. These mutations in LPL and apoCII apparently are not atherogenic, since plasma cholesterol and LDL levels are very low in these patients.

Hyper-VLDL, also called Fredrickson Type IV, is caused by an overproduction of VLDL and/or a reduced catabolism. The exact genetic defect of Type IV is still under investigation.

3.4.3 Mixed Hyperlipidemias Including Familial Dys-β-lipoproteinemia and Familial Combined Hyperlipidemia

Here it is not always possible to strictly dissect primary from secondary forms. The classical mixed HLP is Type V according to Fredrickson where chylomicrons, VLDL, and LDL are elevated in fasting plasma. The exact molecular defect is not known, yet there are several hormones and enzymes involved in the pathophysiology of this disease.

Patients suffering from Type IIB hyperlipoproteinemia exhibit elevated plasma levels of LDL (β-Lp) and VLDL (pre-β-Lp) of unknown etiology. This form of HLP is also strongly atherogenic.

Familial combined HLP has been denominated a multigenic disorder where hypercholesterolemia, hypertriglyceridemia, and mixed forms occur simultaneously within one family.

Familial Dysbetalipoproteinemia, Type III According to Fredrickson: The genetic defect of Type III HLP has been clarified by Utermann (1988) who recognized a mutation in apoE of these patients. ApoE that binds to the LDL-R in addition to LRP exists in three major polymorphic alleles, APOE2, APOE3, and APOE4, where APOE3 is the wild type. ApoE2 shows a strongly reduced receptor affinity and gives rise to an IDL-like broad β-Lp particle with altered density and electrophoretic mobility, i.e., dys-β-Lp. ApoE2 by itself does not cause hyperlipidemia. Only if additional disorders are present such as thyroid dysfunction, the classical Type III HLP develops. Type III HLP is very atherogenic and demands treatment, where fibrates are the drugs of choice. Individuals with APOE2 are protected from Alzheimer's disease (AD), whereas APOE4 has been recognized as a strong risk factor for maturity onset AD.

3.4.4 Hypo-HDL

The most striking hypo-HDL is Tangier disease (TD), called according to the island Tangier where the first family with absent plasma HDL was observed. It took centuries to unravel the genetic defect of TD, since everyone was searching for defects in the gene of apoAI. Finally, the defect in TG was recognized as mutation in the ABCA-1 transporter. In this disorder, nascent HDL do not mature due to the lack of cholesterol efflux from peripheral cells and are very fast catabolized in the kidney. There are numerous mutations and polymorphisms known in ABCA-1 that occur in heterozygous, homozygous, and compound heterozygous forms, leading to more or less reduced plasma HDL concentrations.

Nascent HDL on the other hand are also biosynthesized during lipolysis of chylomicrons and VLDL in the form of surface remnants. Thus, any derangement in the lipolytic process of TG-rich lipoproteins is accompanied by low plasma HDL levels. Hypo-HDL is also a consequence of excess activity of CETP and EDL.

Whether or not low plasma HDL is causally linked to the development of atherosclerosis or might represent only a risk indicator is currently under heavy debate. Most probably what matters is not so much the plasma level but rather the turnover of HDL that influences atherogenesis.

3.4.5 Hyper-Lp(a)

Lp(a) belongs to the class of CE-rich apoB-containing lipoproteins. Originally Lp(a) was believed to represent a genetic isoform of LDL. Later it was recognized that Lp(a) consists of an LDL-like core particle with apo(a), a large glycoprotein, attached to apoB-100 via a disulfide bridge (for a review see Kostner and Kostner 2004). Despite of intensive research over the last decades, the physiological function of Lp(a) is merely unknown. The striking feature of apo(a) is its great homology to plasminogen with the so-called kringles as the most characteristic structural elements. These kringles are repeated in apo(a) several fold at variable numbers and give rise to a characteristic size polymorphism. In fact, apo(a) size varies from approx. 200–600 kD, and it was found that apo(a) size exhibit a strong and significant negative correlation with plasma Lp(a) levels.

3.4.5.1 Primary Hyper-Lp(a)

Turnover studies in man established that under healthy conditions Lp(a) concentrations are determined by the rate of apo(a) biosynthesis that is under genetic control. Lp(a) levels in fact are not normally distributed, which is best realized by the fact that mean values in our population are approx. 10 mg/dl and median values 20 mg/dl. Lp(a) appears to be one of the most atherogenic Lp. The mechanism by which it exerts its atherogenicity is not fully explored, yet there exist several hypotheses that may act together. Due to the homology with plasminogen, apo(a) competes with fibrin binding and fibrinolysis. Lp(a) also increases the expression of PAI-I and displaces plasmin from binding to fibrin clots. Lp(a) finally interferes with the activity of metalloproteinases and the conversion of pre-TGF-β to its mature form.

Recently published large epidemiological studies demonstrate beyond any doubt that increased plasma Lp(a) concentrations are causally related to cardiovascular disease and myocardial infarction (Clarke et al. 2009). There was also a consensus report published by the European Atherosclerosis Society; the cutoff level for plasma Lp(a) was postulated to be 50 mg/dl, above which a serious additional cardiovascular risk has to be considered. Unfortunately there are only few medications known that may reduce plasma Lp(a). Conventional lipid-lowering drugs including statins, fibrates, and bile acid sequestrants have a variable effect and in some cases even lead to a significant increase of Lp(a). The only drug that shows a more consistent effect is nicotinic acid and its derivatives (reviewed by Kostner and Kostner 2005). Nicotinic acid, however, has unwanted adverse effects and is therefore registered only in few countries. Recent studies of our institute, however, are promising that FXR agonists that are currently in clinical trials such as INT-747 may become a very effective class of medications affecting increased Lp(a) levels. FXR is a nuclear receptor that regulates directly or indirectly the activity of numerous lipogenic proteins including CYP-7A1; apoE; apoCI, II, and III; PLTP; SR-BI; VLDL-R; and SREBP-1c but also genes involved in glucose metabolism (reviewed in Modica et al. 2009). In our studies we have demonstrated that natural and synthetic FXR ligands profoundly downregulate apo(a) expression. This was due to a direct effect by competitively inhibiting the binding of HNF4α, a master regulator of genes expressed in the liver, and in addition due to the induction of FGF19/15 biosynthesis in the intestine that signals via the FGFR4 to the liver and represses apo(a) expression.

3.4.5.2 Secondary Hyper-Lp(a)

Plasma Lp(a) levels are stable over time and are hardly influenced by diet and drugs. There are, however, diseases that profoundly influence plasma Lp(a). It has been found that patients with nephrotic syndrome have 3- to 5-fold elevated Lp(a) which probably is due to an increased apo(a) biosynthesis. Patients with end-stage renal disease on the other hand show a reduced catabolism of Lp(a) and 2- to 3-fold elevations. It is believed that these Lp(a) elevations contribute significantly to the atherogenicity of kidney disease.

Significantly elevated plasma Lp(a) levels are also found in FH patients, yet the molecular mechanism of this elevation is unknown. In a similar way, hypothyroid patients show significantly elevated Lp(a) that are "normalized" under appropriate treatment of the primary disease.

3.4.6 Other Dyslipoproteinemias

Under normal fasting conditions, human plasma contains defined Lp density classes that are characterized in Chap. 1 in Vol. I. Any genetic predisposition may alter this distribution, giving rise to the appearance of "dys-Lp" that are either detectable only in postprandial plasma or may represent a nonphysiological fraction. Examples are HLP Type III or Type V HLP.

3.4.6.1 Lipoprotein-X (LP-X)

LP-X belongs to the class of LDL, yet it contains no apoB but consists mostly of apoC in addition to FC and PL. LP-X is absent in healthy individuals but is found in large concentrations in cholestatic patients (Seidel 1987). For a long time LP-X was used as a clinical-chemical parameter to distinguish between extra- and intrahepatic cholestasis. Since there are no large-throughput methods for LP-X quantitation, this parameter is not routinely used anymore. LP-X is not atherogenic and also found in individuals with primary or secondary LCAT deficiency.

3.5 Hyperlipidemias Associated with Hormonal Disorders

3.5.1 The HLP Triad: Type II Diabetes Mellitus (T2DM), Metabolic Syndrome (MetS), and Hepato-Steatosis

More information on this topic is found in Chap. 2 of this book, and thus, we will give here only a short summary. A characteristic feature of the HLP triad is impaired glucose tolerance and reduced insulin sensitivity with hyperglycemia. The lipoprotein pattern in T2DM, MetS, and fatty liver is rather similar and characterized by an excess flux of free fatty acids to the liver as schematically displayed in Fig. 3.7. The increased flux of FFA leads to an overproduction and a reduced catabolism of TG-rich lipoproteins and Fredrickson Type IIB, IV, and V phenotypes. In addition, sd LDL that are prone to oxidation and thus exhibit a high peroxide content, absence of HDL2 and low HDL3 values, and deficiencies of pre-β-HDL have been reported. The particular phenotype that is observed highly depends on the individual genetic background in addition to environmental factors including diet. This multifactorial disease is probably a result of inappropriate signaling pathways related to nuclear receptor function including among others PPAR-γ, FXR, and LXR. The lipoprotein pattern under these conditions is very atherogenic and thus considered to be one of the major risk factors for atherosclerosis and CHD. In contrast to T2DM,

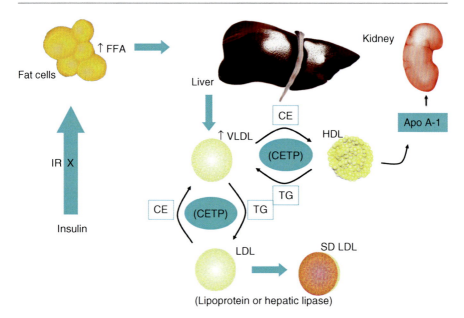

Fig. 3.7 Mechanisms of insulin resistance and dyslipidemia

insulin-dependent diabetes mellitus is characterized by different abnormalities of lipid metabolism and of hemostasis that causes microvascular diseases.

3.5.2 Dys- and Hyper-/Hypolipoproteinemias in Other Hormonal Disorders

In addition to insulin, thyroid and steroid hormones have profound impact on the plasma lipid and lipoprotein metabolism. T3, for example, has been shown to regulate cholesterol biosynthesis and LDL-R activity. Thus, in hypothyroid patients Type IIA- and Type IIB-like phenotypes are observed. In patients with the apoE2 allele, Type III HLP with broad β-Lp bands occurs. The opposite might be the case in hyperthyroid patients.

Male sex hormones and anabolic steroids increase cholesterol biosynthesis, leading to high LDL and low HDL plasma concentrations. This is caused by the interference with the interplay of the nuclear receptors LXR, FXR, VDR, and CAR. Estrogens on the other hand stimulate TG synthesis and may cause elevated plasma VLDL but also HDL concentrations. The whole metabolic effects of sex hormones are complex and very much depend on other secondary factors and genetic dispositions. Simplifying matters we may generalize that androgens lead to an atherogenic Lp pattern but estrogens to an anti-atherogenic – yet thrombogenic – stage.

3.6 Epidemiology of Hyperlipidemias

The negative impact of dyslipidemia is evident from the epidemiologic, angiographic, and *postmortem* studies that have acknowledged the fundamental relationship between elevated serum cholesterol levels and the development of coronary heart diseases. Especially in the presence of comorbid conditions such as diabetes, smoking, and/or hypertension, dyslipidemia is a prominent risk factor for micro- and macrovascular complications. Numerous epidemiological studies in different populations have been published; among them are the classical Framingham study in the USA (Wilson 1994) and the PROCAM (Assmann et al. 2004) and LURIC study (Winkelmann et al. 2001) in Europe. At the basis of these studies, guidelines have been recommended such as the NCEP ATP-III guidelines that served as a model for numerous national guidelines for the assessment of cutoff levels for plasma lipoprotein concentrations and treatment recommendations (Talbert 2003). In all these studies, high levels of LDL cholesterol as well as low HDL cholesterol came out as significant risk factors for CHD. Additionally, dyslipidemia is recognized as an important component in the definition of metabolic syndrome by the International Diabetes Federation (IDF) criteria (2005).

Hypertriglyceridemia has also been reported as an independent risk factor for CHD, directly or indirectly in association with other risk factors including obesity, metabolic syndrome, proinflammatory and prothrombotic biomarkers, and type 2 diabetes mellitus (Grundy et al. 2004; Yuan et al. 2007). Very high levels of TG (>5 mmol/l) also increases the risk of acute pancreatitis. Some triglyceride-rich lipoproteins (TGRLP), the remnant lipoproteins including small- (VLDL) and intermediate-density lipoproteins (IDL), are also considered atherogenic (Grundy 2002). VLDL cholesterol is the usual marker for remnant lipoproteins and is a potential secondary target of cholesterol-lowering therapy. Although LDL-C is identified as the primary target for drug therapy, the current diagnostic trend is to identify triglycerides and apolipoprotein B.

3.7 Dys- and Hyperlipidemias as Risk for Coronary Heart Disease

There is no doubt that apoB-containing lipoproteins including Lp(a), LDL (Ross 1999), β-VLDL, and chylomicron remnants are the most atherogenic lipoproteins. There are numerous concepts explaining the mechanism of action related to atherogenesis. A key feature in atherogenesis are inflammatory self-perpetuating stimuli by cytokines and growth factors brought about by the interaction of monocytes and macrophages with oxidized, glycated, or abnormally modified apoB-containing Lp (Fig. 3.8). At high plasma LDL concentrations, for example, in patients with FH, there is an increased influx of these Lp into the subendothelial layer. Since apoB-containing Lp in general have a high affinity to heparan sulfate proteoglycans that

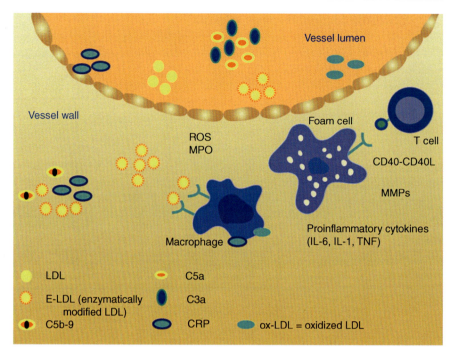

Fig. 3.8 Role of lipoproteins in atherogenesis. *ROS* reactive oxygen species, *MPO* myeloperoxidase, *MMP* metalloproteinase

reside in the arterial intima, they give rise to fatty streak formation. Fatty streaks are reversible – yet only at low plasma cholesterol concentrations. In untreated D&HLP patients, these Lp increase in concentration and attract monocytes, T lymphocytes, and immune cells. These cells are transformed into foam cells and synthesize and secrete chemoattractants, interleukins, and growth factors, leading to local inflammation by a self-perpetuating mechanism. Cytokines cause smooth muscle cell proliferation and transformation from the elastic subtype into the synthetic type and in turn give rise to large amounts of connective tissue, promoting the formation of atherosclerotic plaques.

HDL in contrast are considered to be anti-atherogenic. The mechanisms involved in the anti-atherogenic processes relate in first instance to their role in cholesterol efflux and mobilization of cholesterol from foam cells followed by RCT. In addition, HDL are anti-inflammatory and protect apoB-containing Lp from oxidation. An important enzyme with antioxidative capability is the paraoxonase (PON-1), which neutralizes oxidized lipids. PON-1 expression has been shown to correlate significantly with plasma HDL-C levels. There are, however, also other mechanisms ascribed to HDL that protect from atherogenesis.

Depending on their composition, atherosclerotic plaques may be considered "stable" if they have a thick fibrous cap. Foam cells and lipid-rich plaques with a thin cap on the other hand are considered to be unstable or "vulnerable." If such plaque rupture occurs, subendothelial material, mainly collagen and lipid-loaded macrophages, gets in direct contact with blood components, leading to platelet aggregation and clot formation. These latter events are hallmarks of arterial occlusions and myocardial infarction.

3.8 Perspectives

The years from 1950 to 1980 were characterized for intensive research in the field of lipoprotein biochemistry and structure, methods in clinical chemistry, and metabolism. During this time, the Nobel Prize was granted to the highly distinguished scientists in this field, Joseph Goldstein and Michael Brown, for the detection of the LDL receptor. In the following years, further pioneering work has been published from this laboratory, such as the detection of the role of ABC transporters in cholesterol metabolism, the role of PCSK9 in LDL-receptor abundance, and genes involved in metabolic syndrome. This was only possible by studies involving genetic epidemiology in combination with bioinformatics that covered the years of 1990–2000. From there on, the research in this field was dominated by genomics, proteomics, lipidomics, and metabolomics leading to systems biology. Contemporary methodologies, such as ultrafast sequencing, DNA and RNA arrays, and other high-throughput methodology, particularly in diagnosing and investigating hyper- and dyslipoproteinemias and factors involved in the pathogenesis of atherosclerotic diseases, will certainly lead to fast progress in solving numerous open questions related to the pathophysiology of lipid and lipoprotein metabolism. Without doubt, with such instruments in hand, it will be also possible to make fast progress in developing new medications for treatment of lipid disorders.

References

Alaupovic P, Fruchart JC, Puchois P (1985) Focus on the classification of plasma lipoproteins. Ann Biol Clin (Paris) 43:831–840

Assmann G, Nofer JR, Schulte H (2004) Cardiovascular risk assessment in metabolic syndrome: view from PROCAM. Endocrinol Metab Clin North Am 33:377–392

Bachorik PS, Albers JJ (1986) Precipitation methods for quantification of lipoproteins. Methods Enzymol 129:78–100

Boren J, Taskinen MR, Adiels M (2012) Kinetic studies to investigate lipoprotein metabolism. J Intern Med 271:166–173

Brunzell JD, Chait A, Bierman EL (1978) Pathophysiology of lipoprotein transport. Metabolism 27:1109–1127

Choi SH, Ginsberg HN (2011) Increased very low density lipoprotein (VLDL) secretion, hepatic steatosis, and insulin resistance. Trends Endocrinol Metab 22:353–363

Clarke R, Peden JF, Hopewell JC, Kyriakou T, Goel A, Heath SC, Parish S, Barlera S, Franzosi MG, Rust S, Bennett D, Silveira A, Malarstig A, Green FR, Lathrop M, Gigante B, Leander K, de Faire U, Seedorf U, Hamsten A, Collins R, Watkins H, Farrall M, PROCARDIS Consortium (2009) Genetic variants associated with Lp(a) lipoprotein level and coronary disease. N Engl J Med 361:2518–2528

Courtney JJ, Janssen I (2006) Distribution of lipoproteins by age and gender in adolescents. Circulation 114:1056–1062

Dolphin PJ (1985) Lipoprotein metabolism and the role of apolipoproteins as metabolic programmers. Can J Biochem Cell Biol 6:850–869

Dominiczak MH, Caslake MJ (2011) Apolipoproteins: metabolic role and clinical biochemistry applications. Ann Clin Biochem 48:498–515

Fredrickson DS, Levy RI, Lees RS (1967) Fat transport in lipoproteins–an integrated approach to mechanisms and disorders. N Engl J Med 276:94–103

Goldstein JL, Brown MS (2009) The LDL receptor. Arterioscler Thromb Vasc Biol 29:431–438

Gordon SM, Deng J, Lu LJ, Davidson WS (2010) Proteomic characterization of human plasma high density lipoprotein fractionated by gel filtration chromatography. J Proteome Res 9:5239–5249

Grundy SM (2002) Low-density lipoprotein, non-high-density lipoprotein, and apolipoprotein B as targets of lipid-lowering therapy. Circulation 106:2526–2529

Grundy SM, Cleeman JI, Merz CN (2004) Implications of recent clinical trials for the National Cholesterol Education Program Adult Treatment Panel III guidelines. Circulation 110:227–239

Havel RJ (1995) Chylomicron remnants: hepatic receptors and metabolism. Curr Opin Lipidol 6:312–316

Hazzard WR, Goldstein JL, Schrott MG, Motulsky AG, Bierman EL (1973) Hyperlipidemia in coronary heart disease. 3. Evaluation of lipoprotein phenotypes of 156 genetically defined survivors of myocardial infarction. J Clin Invest 52:1569–1577

Herz J, Willnow TE (1995) Lipoprotein and receptor interactions in vivo. Curr Opin Lipidol 6:97–103

Hoofnagle AN, Heinecke JW (2009) Lipoproteomics: using mass spectrometry-based proteomics to explore the assembly, structure, and function of lipoproteins. J Lipid Res 50:1967–1975

Horton JD, Cohen JC, Hobbs HH (2009) PCSK9: a convertase that coordinates LDL catabolism. J Lipid Res 50(Suppl):S172–S177

International Diabetes Federation (2005) The IDF consensus worldwide definition of the metabolic syndrome. Available at: http://www.idf.org/webdata/docs/IDF_Meta_def_final.pdf. Accessed 1 May 2009

Kolovou GD, Kostakou PM, Anagnostopoulou KK (2011) Familial hypercholesterolemia and triglyceride metabolism. Int J Cardiol 147:349–358

Kontush A, Chapman MJ (2010) Lipidomics as a tool for the study of lipoprotein metabolism. Curr Atheroscler Rep 12:194–201

Kostner GM (1983) Apolipoproteins and lipoproteins of human plasma: significance in health and in disease. Adv Lipid Res 20:1–43

Kostner KM, Kostner GM (2004) Factors affecting plasma lipoprotein(a) levels: role of hormones and other nongenetic factors. Semin Vasc Med 4(2):211–214

Kostner KM, Kostner GM (2005) Therapy of hyper-Lp(a). Handb Exp Pharmacol 2005:519–536

Kostner GM, Oettl K, Jauhiainen M, Ehnholm C, Esterbauer H, Dieplinger H (1995) Human plasma phospholipid transfer protein accelerates exchange/transfer of alpha-tocopherol between lipoproteins and cells. Biochem J 305:659–667

Linton MF, Fazio S (1999) Macrophages, lipoprotein metabolism, and atherosclerosis: insights from murine bone marrow transplantation studies. Curr Opin Lipidol 10:97–105

Lu J, Mitra S, Wang X, Khaidakov M, Mehta JL (2011) Oxidative stress and lectin-like ox-LDL-receptor LOX-1 in atherogenesis and tumorigenesis. Antioxid Redox Signal 15:2301–2333

Modica S, Bellafante E, Moschetta A (2009) Master regulation of bile acid and xenobiotic metabolism via the FXR, PXR and CAR trio. Front Biosci 14:4719–4745

Myers GL, Cooper GR, Sampson EJ (1984) Traditional lipoprotein profile: clinical utility, performance requirement, and standardization. Atherosclerosis 108(Suppl):S157–S169

Nakajima K, Nakano T, Tokita Y, Nagamine T, Inazu A, Kobayashi J, Mabuchi H, Stanhope KL, Havel PJ, Okazaki M, Ai M, Tanaka A (2011) Postprandial lipoprotein metabolism: VLDL vs. chylomicrons. Clin Chim Acta 412:1306–1318

Olofsson SO, Asp L, Boren J (1999) The assembly and secretion of apolipoprotein B-containing lipoproteins. Curr Opin Lipidol 10:341–346

Olofsson SO, Wiklund O, Boren J (2007) Apolipoproteins A-I and B: biosynthesis, role in the development of atherosclerosis and targets for intervention against cardiovascular disease. Vasc Health Risk Manag 3:491–502

Pownall HJ, Morrisett JD, Sparrow JT, Smith LC, Shepherd J, Jackson RL, Gotto AM Jr (1979) A review of the unique features of HDL apoproteins. Lipids 14:428–434

Rifai N, Warnick GR, McNamara JR, Belcher JD, Grinstead GF, Frantz ID Jr (1992) Measurement of low-density-lipoprotein cholesterol in serum: a status report. Clin Chem 38:150–160

Ross R (1999) Atherosclerosis – an inflammatory disease. N Engl J Med 340:115–126

Schneider WJ, Nimpf J, Bujo H (1997) Novel members of the low density lipoprotein receptor superfamily and their potential roles in lipid metabolism. Curr Opin Lipidol 8:315–319

Seidel D (1987) Lipoproteins in liver disease. J Clin Chem Clin Biochem 25:541–551

Sun HY, Chen SF, Lai MD, Chang TT, Chen TL, Li PY, Shieh DB, Young KC (2010) Comparative proteomic profiling of plasma very-low-density and low-density lipoproteins. Clin Chim Acta 411:336–344

Tadey T, Purdy WC (1995) Chromatographic techniques for the isolation and purification of lipoproteins. J Chromatogr B Biomed Appl 671:237–253

Talbert RL (2003) Role of the National Cholesterol Education Program Adult treatment panel III guidelines in managing dyslipidemia. National Cholesterol Education Program Adult treatment panel III. Am J Health Syst Pharm 60(13 Suppl 2):3–8

Utermann G (1988) Apolipoprotein polymorphism and multifactorial hyperlipidaemia. J Inherit Metab Dis 11(Suppl 1):74–86

Vrins CL (2010) From blood to gut: direct secretion of cholesterol via transintestinal cholesterol efflux. World J Gastroenterol 16:5953–5957

Warnick GR (1986) Enzymatic methods for quantification of lipoprotein lipids. Methods Enzymol 129:101–123

Warnick GR (1990) Laboratory measurement of lipid and lipoprotein risk factors. Scand J Clin Lab Invest Suppl 198:9–19

Wilson PW (1994) Established risk factors and coronary artery disease: the Framingham study. Am J Hypertens 7(7 Pt 2):7S–12S

Winkelmann BR, Marz W, Boehm BO, Zotz R, Hager J, Hellstern P, Senges J (2001) Rationale and design of the LURIC study – a resource for functional genomics, pharmacogenomics and long-term prognosis of cardiovascular disease. LURIC Study Group (LUdwigshafenRIsk and Cardiovascular Health). Pharmacogenomics 2(1 Suppl 1):S1–S73

Xiao C, Hsieh J, Adeli K, Lewis GF (2011) Gut-liver interaction in triglyceride-rich lipoprotein metabolism. Am J Physiol Endocrinol Metab 301:E429–E446

Yamada Y, Doi T, Hamakubo T, Kodama T (1998) Scavenger receptor family proteins: roles for atherosclerosis, host defence and disorders of the central nervous system. Cell Mol Life Sci 54:628–640

Ye D, Lammers B, Zhao Y, Meurs I, Van Berkel TJ, Van Eck M (2011) ATP-binding cassette transporters A1 and G1, HDL metabolism, cholesterol efflux, and inflammation: important targets for the treatment of atherosclerosis. Curr Drug Targets 12:647–660

Yuan G, Al-Shali KZ, Hegele RA (2007) Hypertriglyceridemia: its etiology, effects and treatment. Can Med Assoc J 176:1113–1120

Hyperuricemia

Tetsuya Yamamoto, Masafumi Kurajoh, and Hidenori Koyama

Abstract
Most cases of hyperuricemia are ascribable to a combination of serum urate level-increasing genetic factors such as ABCG2 mutation and environmental factors including excessive purine and/or ethanol ingestion. Although it is well known that hyperuricemia causes gout, it remains uncertain whether it causes or accelerates lifestyle-related diseases, such as cardiovascular disease (CVD) and chronic kidney disease (CKD). Recent epidemiological studies have indicated that hyperuricemia is a risk factor for cardiovascular diseases, stroke, diabetes mellitus (DM), and CKD, as well as others. Several experimental and clinical studies have strongly suggested that hyperuricemia causes and accelerates CVD and CKD. However, the numbers of subjects in those clinical investigations including intervention studies were not adequate to definitively conclude that hyperuricemia causes CVD and/or CKD. Accordingly, additional examinations are needed, especially in the area of clinical research.

Keywords
Uric acid • Hyperuricemia • Gout • Cardiovascular disease

4.1 Introduction

Hyperuricemia is defined as a concentration of urate in serum over 7 mg/dl, which induces gout and is manifested by attacks of acute arthritis, the so-called gout flare, deposits of monosodium urate monohydrate in and around the joints of the extremities and in subcutaneous tissues, renal disease involving interstitial tissues and blood vessels, and uric acid nephrolithiasis. Thus, the condition is considered

---

T. Yamamoto (✉) • M. Kurajoh • H. Koyama
Division of Endocrine and Metabolism, Department of Internal Medicine,
Hyogo College of Medicine, 1-1 Mukogawa-cho, Nishinomiya, Hyogo 663-8501, Japan
e-mail: tetsuya@hyo-med.ac.jp

to be an important factor for gout development. Hyperuricemia usually develops from a combination of serum urate-increasing genetic and environmental factors such as lifestyle. Accordingly, it is commonly seen in advanced countries and increases together with Westernized changes in lifestyle including diet in developing countries (Johnson et al. 2005a). In addition, the condition is closely related to lifestyle-related diseases including hypertension, chronic kidney disease (CKD), and metabolic syndrome, which accelerate cardiovascular disease (Johnson et al. 2005b), and are also important factors to induce hyperuricemia. On the other hand, recent experimental studies have shown that hyperuricemia induces hypertension, CKD, and metabolic syndrome (Mazzali et al. 2001; Kang et al. 2002; Sánchez-Lozada et al. 2008), which are very intriguing findings. In several studies, uric acid has been found to have a prooxidant effect in a hydrophobic state, whereas it has an antioxidant effect in a hydrophilic state (Abuja 1999; Ames et al. 1981; Kuzkaya et al. 2005; Muraoka and Miura 2003). These effects may accelerate or protect against many diseases. Furthermore, other studies have demonstrated that hyperuricemia protected against progression of diseases including Parkinsonism (Constantinescu and Zetterberg 2011; Alonso et al. 2007; Alvarez-Lario and Macarrón-Vicente 2011). On the other hand, several epidemiological studies have suggested that hyperuricemia is an independent risk factor for cardiovascular disease (CVD) (Niskanen et al. 2004; Fang and Alderman 2000; Alderman et al. 1999). Finally, experimental findings have suggested that uric acid has various effects on many different tissues, such as the vessels, kidneys, adipose tissues, and heart, as well as others (Mazzali et al. 2001; Kang et al. 2002; Sánchez-Lozada et al. 2008; Sautin et al. 2007). In this chapter, we focus on accumulated findings obtained in epidemiological and experimental studies of uric acid.

4.2 Etiology of Hyperuricemia

4.2.1 Hyperuricemia Development Is Related to Many Factors

4.2.1.1 Genetic Factors

Genetic factors play an important role in regulation of the serum concentration of urate, especially genetic factors in the renal handling of uric acid. Hyperuricemia is caused by reduced renal excretion of uric acid in more than 90 % of affected patients. A previous segregation analysis showed that the serum concentration of urate is regulated by interactions between major and modifying genes in addition to environmental factors (Wilk et al. 2000). However, it is also known that even single gene disorders can cause hyperuricemia (Table 4.1). Among those disorders, complete deficiency of hypoxanthine-guanine phosphoribosyl phosphorylase (HGPRT) is well known as Lesch-Nyhan disease, an X-linked recessive disorder that causes central nervous system symptoms, gout, and hyperuricemia. HGPRT is a salvage enzyme that converts hypoxanthine and guanine to inosine-5′-monophosphate (IMP) and guanosine-5′-monophosphate (GMP), respectively, using 5-phosphoribosyl

Table 4.1 Genetic factors that cause hyperuricemia

Disease
Hypoxanthine-guanine phosphoribosyltransferase (HPRT) deficiency (Lesch-Nyhan syndrome, Seegmiller syndrome)
Phosphoribosylpyrophosphate synthetase superactivity
Glycogen storage disease type Ia (glucose 6 phosphatase deficiency) (von Gierke's disease)
Glycogen storage disease type Ib (glucose six phosphate transporter deficiency)
Glycogen storage disease type III (glycogen debranching enzyme deficiency)
Glycogen storage disease type V (muscle glycogen phosphorylase deficiency) (McArdle's disease)
Glycogen storage disease type VII (muscle phosphofructokinase deficiency) (Tarui's disease)
Uromodulin-associated kidney disease (uromodulin mutations) (familial juvenile hyperuricemic nephropathy and medullary cystic kidney disease 2)

Table 4.2 Other genetic variants associated with serum urate

ABCG2: ATP-binding cassette subfamily G member 2
SLC2A9: Solute carrier family 2 (facilitated glucose transporter), member 9
SLC17A1: Solute carrier family 17 (sodium phosphate), member 1
SLC22A11: Solute carrier family 22 (organic anion/urate transporter), member 11
PDZK1: PDZ domain containing 1
GCKR: Glucokinase (hexokinase 4) regulator
LRRC16A: Leucine rich repeat containing 16A
SLC22A12: Solute carrier family 22 (organic anion/urate transporter), member 12
SLC16A9: Solute carrier family 16, member 9 (monocarboxylic acid transporter 9)

1-pyrophosphate (PRPP). Accordingly, deficiency of HGPRT results in overproduction of uric acid (hypoxanthine and guanine → xanthine → uric acid). Familial juvenile hyperuricemic nephropathy is also a rare but well-known autosomal dominant disorder that is characterized by reduced fractional excretion of uric acid, as well as defective transport and apical membrane expression of uromodulin, leading to hyperuricemia and renal failure.

Other genes regulating the renal transport of uric acid are candidates involved in hyperuricemia, and many genes regulating the renal transport of uric acid have recently been discovered in a genome-wide association study (GWAS) (Table 4.2, Fig. 4.1) (Kolz et al. 2009). One of those is ABCG2 gene, whose polymorphisms were found to be associated with hyperuricemia and gout. As compared to the wild type, extrarenal urate excretion is lowered by the ABCG2 transporter of mutation Q141K and the ABCG2 transporter of mutation Q126X (Ichida et al. 2012). In Japan, approximately 10 % of gout patients have genotype combinations, resulting in more than 75 % reduction of ABCG2 function. Although many other transporters may be associated with hyperuricemia, further studies are required to identify whether the mutations of those transporters cause hyperuricemia.

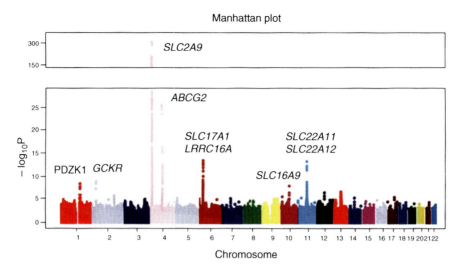

Fig. 4.1 Manhattan plots showing association of all SNPs in meta-analysis of male subjects. SNPs are plotted on the *x*-axis according to their position on each chromosome against association with urate concentration on the *y*-axis (shown as $-\log_{10} p$-value; reprinted from Kolz et al. 2009)

4.2.1.2 Environmental Factors

4.2.1.2.1 Dietary Factors

Purines

Excessive ingestion of meat and seafood is associated with hyperuricemia, as shown in a number of epidemiological studies (Choi et al. 2005). Meat and seafood contain considerable purines, which are converted to uric acid after ingestion. A previous study showed that the daily uric acid turnover rate was 700 mg, which comprises 60 % of the uric acid pool in the body, while approximately 30 % of daily uric acid loss is via excretion through the intestinal tract and the other is excreted via the kidneys (Richards and Weinman 1966; Edwards 2008). Another study demonstrated that ingestion of 1 g of AMP and 1 g of GMP increased the uric acid pool by 1.5- to 2-folds in male subjects and 0.25-fold in females, indicating that purine ingestion increases the uric acid pool, leading to hyperuricemia especially in males. Other studies have found that ingestion of DNA, RNA, purine nucleotides, purine nucleosides, and purine bases increased the serum concentration of urate (Waslien et al. 1968; Clifford et al. 1976) of which RNA, AMP, and adenine increased that concentration to a great degree (Clifford et al. 1976). These results indicate that purines are metabolized differently and produce variable increases in serum concentration of urate, though they are closely related compounds in biochemical term. Nevertheless, all types of purines increase the serum concentration of urate when large amounts are ingested. Accordingly, ingestion of foods containing considerable amounts of purines increases the serum concentration of urate and contributes to hyperuricemia.

4 Hyperuricemia

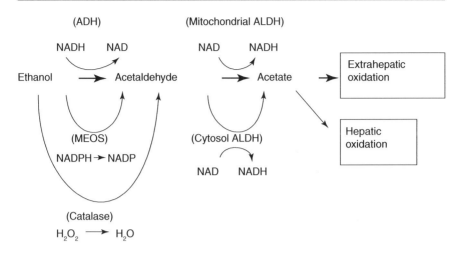

Fig. 4.2 Ethanol elimination pathway. *Bold arrows* show the main pathway. *ADH* alcohol dehydrogenase, *ALDH* aldehyde dehydrogenase, *MEOS* microsomal ethanol-oxidizing system

Ethanol

Light-to-moderate alcohol consumption is associated with reduced mortality from CVD, while heavy drinking is associated with increased CVD (Ruidavets et al. 2010; Ikehara et al. 2008). In addition, heavy drinking increases adenine nucleotide degradation [ATP→ADP→AMP→inosine-5′-monophosphate (IMP)→inosine →hypoxanthine→xanthine→uric acid], leading to a rise in the serum concentrations of urate (Faller and Fox 1982; Puig and Fox 1984). Ethanol is metabolized to acetaldehyde by alcohol dehydrogenase (ADH), the microsomal ethanol-oxidizing system (MEOS) and catalase, then subsequently metabolized to acetate by mitochondrial and cytosolic aldehyde dehydrogenase (ALDH) in the liver (Ugarte and Iturriaga 1976). Acetate formed from ethanol in the liver is transported to peripheral tissues, where it is metabolized to CO_2 and H_2O (Fig. 4.2) (Winkler et al. 1969). During these metabolic processes, adenine nucleotide is degraded by two different mechanisms, increased ATP consumption and decreased ATP production (Puig and Fox 1984; Masson et al. 1992; Yamamoto et al. 2005).

Figure 4.3 presents the mechanism involved in enhanced ATP consumption. In a previous study, the effects of ethanol and acetate infusions were examined in normal subjects following isotopic adenine administration (Puig and Fox 1984). Urinary excretion of oxypurines (hypoxanthine+xanthine) was increased by infusions of adenine, indicating that both ethanol and acetate increase purine nucleotide degradation by accelerating turnover of the adenine nucleotide pool. In addition, those results suggest that acetate formed from ethanol contributes to increased urate production by adenine nucleotide degradation, since acetate is metabolized to acetyl-CoA via acetyl-AMP using ATP, leading to ATP consumption which accelerates adenine nucleotide degradation [ATP→ADP→AMP→IMP→inosine→hypoxanthine→xanthine→uric acid] in peripheral tissues.

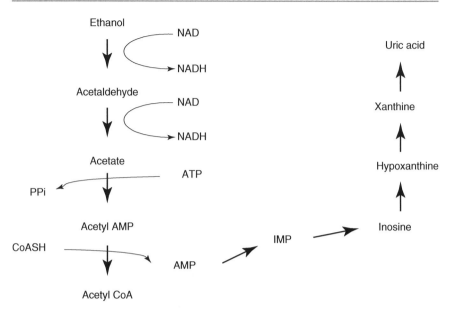

Fig. 4.3 Proposed mechanism for adenine nucleotide degradation during ethanol metabolism. *PPi* inorganic pyrophosphate, *IMP* inosine-5′-monophosphate, *CoASH* coenzyme A

The second mechanism is decreased production of ATP, as shown in Fig. 4.4. Ethanol is oxidized to acetaldehyde by reduction of NAD to NADH mainly by ADH in the cytosol of the liver cells. Subsequently, acetaldehyde is also oxidized to acetate with a reduction of NAD to NADH partly by ALDH in the cytosol and mainly by ALDH in the mitochondria. As a result, the redox potentials of NAD in both the cytosol and mitochondria of hepatocytes are reduced. A previous study reported that ethanol decreased the release of lactic acid and pyruvic acid, increased the concentration of sn-glycerol phosphate, and decreased the concentration of inorganic phosphate in perfused rat liver (Beauvieux et al. 2002). These findings indicate that glycolysis is inhibited during the ethanol-induced reduction in redox potential of NAD in the cytosol of perfused rat hepatocytes. In addition to those, an abrupt decrease in the total hepatic content of nucleoside triphosphate (NTP) that nearly matched ATP loss was noted. Since another study showed that mitochondrial ATP synthesis was increased in ethanol-perfused whole livers (Masson et al. 1992), it is suggested that ATP production in mitochondria in the presence of ethanol is insufficient to alleviate inhibition of glycolytic ATP production in the cytosol. Finally, a decrease in ATP concentration due to diminished ATP production in the cytosol leads to enhanced dephosphorylation of ATP to AMP, resulting in enhanced adenine nucleotide degradation.

An increase in the blood concentration of lactate induced by ethanol metabolism also causes hyperuricemia. In an *in vivo* study using cats, hepatic lactic acid uptake was shown to be suppressed with increased ethanol concentration in the blood (Greenway and Lautt 1990). Accordingly, it is strongly suggested that decreased lactic

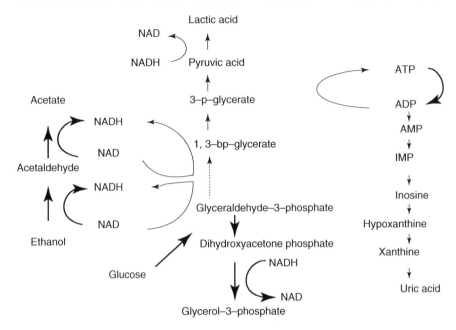

Fig. 4.4 Proposed mechanism for enhanced adenine nucleotide degradation by reduction in redox potentials of nicotinamide adenine nucleotides in the cytosol of liver cells. A reduction in the redox potentials of nicotinamide adenine nucleotides inhibits glycolysis, resulting in enhancement of adenine nucleotide degradation. ⇢ denotes glycolysis inhibition site, *IMP* inosine monophosphate

acid uptake in the liver causes an ethanol-induced increase in the blood concentration of lactate, since ethanol decreased the release of lactic acid and pyruvic acid from perfused rat liver (Masson et al. 1992). An increased blood concentration of lactate also reduces the urinary excretion of uric acid via urate transporter 1 (URAT1) at the proximal tubules, which participates in factors related to ethanol-induced hyperuricemia. Although ethanol itself increases the serum concentration of urate, as described above, purines present in alcohol beverages may also contribute to an increase in the serum concentration of urate. Since beer contains a greater amount of purines than other alcohol beverages, that degradation to uric acid in the body increases the serum concentration of urate. In our study, freeze-dried beer (0.34 g/kg body weight) increased the plasma concentration of urate by approximately 20 μmol/l, while regular beer (10 ml/kg body weight) also increased that by approximately 30 μmol/l (Yamamoto et al. 2002). Since the amount of purines in freeze-dried beer was shown to be the same as in regular beer in that study, those present in beer are indicated to contribute to an increase in plasma concentration of urate.

Fructose
A number of recent epidemiological studies have demonstrated that fructose ingestion is associated with the onset of gout, suggesting that fructose increases the serum concentration of urate (Gao et al. 2007; Choi et al. 2008b, 2010), though

other studies have showed a lack of association between dietary fructose and hyperuricemia (Sun et al. 2010).

Fructose is a simple monosaccharide present in many plants. Since at least half of fructose injected intravenously is metabolized in the liver in humans (Mendeloff and Weichselbaum 1953), its concentration in peripheral blood does not rise markedly after ingesting foods containing fructose or sucrose. However, during its metabolism of fructose in the body, serum urate concentration is increased. In a previous study, intravenously administered fructose increased serum and urinary uric acid concentrations in normal children, with the effect on uric acid shown to be dose dependent and specific (Perheentupa and Raivio 1967).

Fructose-induced hyperuricemia occurs together with a sharp decrease in hepatic ATP concentration and total adenine nucleotides, as well as accumulation of fructose-1-phosphate and phosphate, and decreased inorganic phosphate concentration (Mäenpää et al. 1968; Woods et al. 1970). Taken together, the findings indicate that fructose is rapidly phosphorylated to fructose-1-phosphate by highly active fructokinase, using ATP as a phosphate donor, together with a sharp fall in ATP concentration. As a result, ATP is depleted and ATP loss is more enhanced by inhibition of oxidative phosphorylation of ADP due to a shortage of inorganic phosphate. Furthermore, a shortage of inorganic phosphate relieves allosteric inhibition by inorganic phosphate toward AMP deaminase and accelerates adenine nucleotide degradation (Fig. 4.5).

Recent studies have shown that ingestion of a large amount of fructose is associated with metabolic syndrome, which is closely related with hyperuricemia (Miller and Adeli 2008; Dekker et al. 2010). Accordingly, fructose-induced metabolic syndrome may contribute to an increase in the serum concentration of urate, because metabolic syndrome is strongly suggested to cause hyperuricemia. On the other hand, very intriguing findings from a study using rats showed that a fructose-induced increase in serum concentration of urate caused metabolic syndrome (Nakagawa et al. 2006). Further experimental and clinical studies are needed to clarify the relationship between fructose-induced metabolic syndrome and hyperuricemia.

4.3 Uric Acid: An Independent Risk Factor of CVD

4.3.1 Epidemiological Studies

The serum concentration of urate is associated with hypertension, chronic kidney disease, type 2 diabetes mellitus (DM), metabolic syndrome, and dyslipidemia, all of which are established risk factors of CVD (Johnson et al. 2005a, b; Mazzali et al. 2001; Kang et al. 2002; Sánchez-Lozada et al. 2008; Alper et al. 2005; Dyer et al. 1999; Forman et al. 2007; Hunt et al. 1991; Iseki et al. 2004; Tomita et al. 2000; Nakanishi et al. 2003; Dehghan et al. 2008; Bhole et al. 2010; Choi and Ford 2007; Choi et al. 2008a, b; Emmerson 1998). Since patients with gout are frequently obese middle-aged men complicated with these diseases, CVD is more frequently observed in those patients, as compared with the general population. Nevertheless,

4 Hyperuricemia

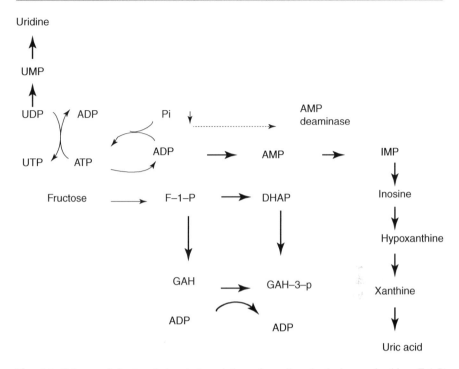

Fig. 4.5 Schema of fructose-induced degradation of uracil and adenine nucleotides. *F-1-P* fructose-1-phosphate, *DHAP* dihydroxyacetone phosphate, *GAH* glyceraldehyde, *GAH-3-P* glyceraldehyde-3-phosphate, *Pi* inorganic phosphate, *UTP* uridine-5′-triphosphate, *UDP* uridine-5′-diphosphate, *UMP* uridine-5′-monophosphate, *IMP* inosine-5′-monophosphate

a number of epidemiological studies have shown that uric acid is an independent risk factor of CVD. A representative study is the First National Health and Nutrition Examination Survey (NHANES I) Epidemiologic Follow-up Study (Freedman et al. 1995), in which baseline data were collected from 5,421 persons from 1971 to 1975 and follow-up data were obtained from 1971 to 1987 in order to examine the association of serum uric acid levels with cardiovascular mortality. Those findings demonstrated that baseline serum urate concentration is an independent predictor of mortality from CVD in both men and women. In addition, to further examine the association of serum urate concentration with CVD, the NHANES study was extended from 1987 to 1992. Those results suggested that increased serum urate concentration is independently and significantly associated with risk of cardiovascular mortality (Fang and Alderman 2000). Furthermore, a population-based study using elderly individuals has shown that the serum urate concentration is associated with myocardial infarction and stroke (Bos et al. 2006). Also, a systemic review and meta-analysis of 238,449 adults indicated that hyperuricemia may modestly increase the risk of both stroke incidence and mortality (Kim et al. 2010).

On the other hand, other epidemiological studies, including the Framingham Heart Study (Culleton et al. 1999), have shown that uric acid is not an independent risk factor. In that study, baseline data were collected from 6,763 individuals from 1971 to 1976 with a total follow-up period of 117,376 person-years. Their findings indicated that uric acid does not have a causal role in the development of CHD, death from cardiovascular disease, or death from all causes. Accordingly, though it is conceivable that uric acid is an independent risk factor of CVD, it remains uncertain.

4.3.2 Experimental Studies

In addition to epidemiological studies, recent experimental findings indicate that uric acid causes arteriosclerosis. A recent study reported that URAT1 mRNA and protein were observed in both non-stimulated and uric acid-stimulated human vascular smooth muscle cells (VSMCs) (Price et al. 2006). In addition, it was shown that uric acid enters human VSMCs and human umbilical vein endothelial cells (HUVECs) via organic acid transporters, presumably URAT1, since probenecid, an inhibitor of uric acid transport, decreases uric acid-induced production of reactive oxygen species (ROS) and subsequent intracellular reactions (Yu et al. 2010). Accordingly, URAT1 seems to be an important transporter of uric acid in VSMCs. Another study showed that apocynin, an NADPH oxidase inhibitor, inhibited cell proliferation and endothelin-1 (ET-1) secretion induced by uric acid in human aortic smooth muscle cells (Chao et al. 2008). In addition, they found that p47phox (a subunit of NADPH oxidase) small interfering RNA knockdown inhibited cell proliferation and ET-1 secretion, strongly suggesting that uric acid activates NADPH oxidase, which produces ROS, and also accelerates cell proliferation and ET-1 secretion via ROS-induced p38 MAPK activation (Chao et al. 2008). Since ET-1, which is associated with atherogenesis as well as cell proliferation, is a cause of atherogenesis, these results suggest that uric acid may contribute to atherogenesis. In another study, uric acid stimulated VSMC DNA synthesis in a concentration-dependent manner and increased the cell number; thus uric acid stimulated VSMC DNA and proliferation (Rao et al. 1991). In addition, expression of platelet-derived growth factor (PDGF) A-chain mRNA appeared after 6 h in uric acid-stimulated VSMCs, peaked at 8 h, and returned to the baseline at 10 h. Also, secretion of PDGF A-chain protein in uric acid-stimulated VSMCs was increased by 10-fold or greater over the control and the anti-PDGF A-chain antibody markedly decreased uric acid-stimulated VSMC [^{3}H]thymidine incorporation. Accordingly, it was suggested that uric acid stimulates VSMC proliferation by induction of the PDGF A-chain, leading to atherogenesis. Furthermore, another study showed that monocyte chemoattractant protein-1 (MCP-1) mRNA expression in VSMCs appeared at 6 h and peaked at 24 h after beginning VSMC incubation in a uric acid-containing media, though MCP-1 mRNA was not generally detectable in VSMCs at the baseline (Kanellis et al. 2003). These results show that MCP-1 protein is secreted from VSMCs after incubation in the uric acid-containing media, and the effects of uric acid on MCP-1 secretion are dose dependent. In addition, uric acid activated transcription factors (NFκB and AP-1) after 15 min of incubation with uric acid, and their activities peaked at 30 min.

Since NFκB and AP-1 have been reported to be involved in regulation of MCP-1 expression, these findings indicate that both transcription factors are activated by uric acid to play a role in uric acid-induced MCP-1 protein production. Furthermore, uric acid (5 mg/dl) activated both ERK1/2 MAPK and p38 MAPK after 15 min in rat aortic VSMCs, while PD 98059 (ERK1/2 MAPK pathway inhibitor) and SB 203580 (p38 MAPK pathway inhibitor) had similar effects on uric acid-induced MCP-1 secretion, decreasing MCP-1-secreted protein by approximately 50 %. Accordingly, these findings indicate that both ERK1/2 MAPK and p38 MAPK play roles in uric acid-induced MCP-1 protein production. Activation of p38 MAPK, ERK1/2 MAPK, and NFκB increases cyclooxygenase-2 (COX-2) production. In VSMCs, uric acid increased COX-2 mRNA expression during 3 and 6 h incubations, while a COX-2 inhibitor (NS398) significantly decreased the uric acid-induced MCP-1-secreted protein, suggesting that COX-2 partially mediates the uric acid-induced increase in MCP-1 synthesis. In addition, COX-2 is thought to mediate uric acid-induced VSMC proliferation in part through thromboxane generation, since COX-2 enhanced VSMC proliferation (Kang et al. 2002). Also, since antiprooxidants (NAC and DPI) inhibited uric acid-induced MCP-1 production by approximately 50 %, uric acid functions in a prooxidant manner and activates NADPH oxidase, resulting in ROS production. Taken together, these findings suggest that uric acid is transported into VSMCs via URAT1 and activates NADPH oxidase. Then, NADPH oxidase-induced ROS production activates MAPK, followed by transcription factor (NFκB and AP-1)- induced synthesis of COX-2 and ET-1, and finally PDGF and MCP-1 via COX-2 activation (Fig. 4.6). Since PDGF accelerates vascular smooth muscle proliferation and MCP-1 induces inflammation, these factors may contribute to atherogenesis.

In HUVECs, it was also shown that uric acid increased ROS production at concentrations of 6 mg/dl or higher after 15 min of uric acid incubation, decreased serum-induced proliferation of HUVECs at concentrations of 6 mg/dl or higher after 48 h of uric acid incubation, and increased senescence and apoptosis of HUVECs at concentrations of 9 mg/dl or higher after 48 h of uric acid incubation, while uric acid-induced apoptosis and senescence were ameliorated by the antioxidants, NAC and tempol (Yu et al. 2010). In addition, that study found that uric acid increased local mRNA expression of angiotensinogen, angiotensin-converting enzyme (ACE), angiotensin II type 1 receptor, and angiotensin II type 2 receptor in HUVECs after 1 h of uric acid incubation (Yu et al. 2010). Furthermore, uric acid increased the production of angiotensin II in cell lysate at 24 h after beginning uric acid incubation, which was inhibited by probenecid and tempol. Uric acid-induced alterations in cell proliferation, senescence, and apoptosis of HUVECs were also inhibited by probenecid and RAS blockers such as enalaprilat and telmisartan at 3 h after beginning uric acid incubation. However, neither of those RAS blockers inhibited ROS production at 30 min after beginning uric acid incubation, while probenecid did (Yu et al. 2010). Taken together, these findings suggest that uric acid produces ROS (Fig. 4.6), which increase the mRNA expression of angiotensin and ACE, leading to enhanced angiotensin II production.

It is also suggested that increased angiotensin II produces ROS, decreases cell proliferation, and increases senescence and apoptosis of HUVECs. Furthermore,

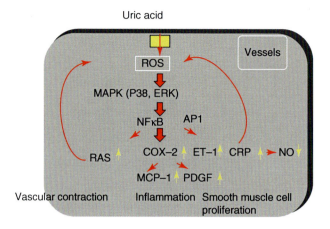

Fig. 4.6 Proposed mechanism of uric acid-induced arteriosclerosis. *ROS* reactive oxygen species, *MAPK* mitogen-activated protein kinase, *ERK* extracellular signal-regulated kinase, *NFκB* nuclear factor kappa-light-chain-enhancer of activated B cells, *AP-1* activator protein-1, *COX-2* cyclooxygenase-2, *RAS* renin-angiotensin system, *ET-1* endothelin-1, *CRP* C-reactive protein, *NO* nitric oxide, *MCP-1* monocyte chemoattractant protein-1, *PDGF* platelet-derived growth factor

those findings suggest that the RAA system plays a role in uric acid-induced ROS production partly. In another study, uric acid enhanced C-reactive protein (CRP) expression in human smooth muscle cells and endothelial cells, which was inhibited by probenecid (Kang et al. 2005a). Also, Kobayashi et al. (2003) reported that the expression of vascular CRP is closely associated with NAD(P)H oxidase, an important enzymatic producer of reactive oxygen species (ROS) in the vessel walls. In another study, CRP directly upregulated NAD(P)H oxidase p22phox and enhanced ROS generation in cultured coronary artery smooth muscle cells (Woods et al. 1970). Thus, CRP may be a direct participant in vascular inflammation and lesion formation via its potent biological effects.

It was recently demonstrated that CRP impairs endothelial NO synthase (eNOS)-dependent vasodilation and uncouples eNOS *in vivo*, suggesting that elevated CRP levels are associated with endothelial dysfunction (Fig. 4.6) (Hein et al. 2009). Uric acid also impaired NO production in bovine aortic endothelium and decreased NO production in cultured pulmonary artery endothelial cells (Khosla et al. 2005; Zharikov et al. 2008). Thus, uric acid is suggested to have widespread effects by inhibiting vascular endothelial NO production. It was also reported that pulmonary arterial endothelial NO synthase activity was not changed by uric acid and pulmonary arterial endothelial L-arginase activity was increased by uric acid without alterations in arginine uptake, suggesting that uric acid-induced arginase activation is an important factor for reduction of NO production, which inhibits dilatation of vessels. Accordingly, the inhibitory effect of uric acid on NO production may result in acceleration of arteriosclerosis.

These experimental results strongly suggest that uric acid accelerates arteriosclerosis and contributes to an increase in patients complicated with CVD.

4.3.3 Relationship Between Hyperuricemia and Risk Factors for CVD

Atherosclerosis-enhancing diseases, such as metabolic syndrome and hypertension, are frequently associated with hyperuricemia and suggested to cause hyperuricemia. On the other hand, hyperuricemia has also been suggested to cause the development of these diseases, resulting in acceleration of atherosclerosis via their development (Johnson et al. 2005b; Sánchez-Lozada et al. 2008; Sautin et al. 2007).

4.3.3.1 Metabolic Syndrome

Metabolic syndrome is a constellation of interrelated metabolic risk factors of CVD which consists of hypertension, hyperglycemia, dyslipidemia, and obesity and is a strong risk factor for cardiovascular disease. Several committees have made a variety of criteria for diagnosis of metabolic syndrome that are applicable for clinical practice (Di Chiara et al. 2012). Although the underlying pathophysiology has not been fully clarified, insulin resistance is considered to be a key factor for the occurrence of metabolic syndrome. Furthermore, visceral fat accumulation plays an important role in insulin resistance. Accordingly, visceral fat accumulation even in mildly obese subjects is associated with the occurrence of cardiovascular disease.

A number of epidemiological studies have demonstrated that metabolic syndrome is associated with hyperuricemia (Choi and Ford 2007; Lee et al. 2011; Chiou et al. 2010; Inokuchi et al. 2010). Another previous study found that the visceral fat area was positively correlated with the plasma concentration of urate and 24 h urinary excretion of uric acid and was negatively correlated with uric acid clearance in healthy subjects (Takahashi et al. 1997). In addition, a different study showed that the visceral fat area was positively correlated with homeostasis model assessment as an index of insulin resistance (HOMA-IR) in both healthy subjects and gout patients (Takahashi et al. 2001), while insulin resistance was inversely correlated with uric acid clearance (Takahashi et al. 2001). These results suggest that insulin resistance due to increased visceral fat causes a decrease in uric acid clearance, resulting in hyperuricemia. Insulin resistance was also reported to be inversely correlated with uric acid clearance (Facchini et al. 1991). In a previous hyperinsulinemic euglycemic clamp study, insulin decreased the renal excretion rates of sodium and uric acid, and the decrease in renal excretion of sodium was significantly correlated with that of uric acid, suggesting that insulin activates reabsorption of sodium and uric acid in the kidneys (Ter Maaten et al. 1997).

Several different studies have shown that insulin activates a variety of transporters that reabsorb sodium at the proximal tubules, suggesting that insulin activates sodium-dependent monocarboxylate transporter 1 (SMCT1) and accelerates reabsorption of sodium and anion (Quiñones-Galvan and Ferrannini 1997). Enhanced reabsorption of anion accelerates both the excretion of anion and reabsorption of uric acid via urate transporter 1 (URAT1), resulting in an increase in reabsorption of uric acid coupled with that of sodium (Anzai et al. 2007). Accordingly, it is suggested that visceral fat obesity-induced insulin resistance leads to hyperinsulinemia, which stimulates reabsorption of uric acid via URAT1, resulting in an increase in

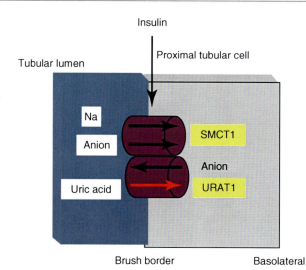

Fig. 4.7 Proposed mechanism of insulin-induced hyperuricemia. Insulin accelerates reabsorption of sodium and anion via SMCT1, followed by enhanced reabsorption of uric acid and excretion of anion via URAT1. *SMCT1* sodium monocarboxylate transporter 1, *URAT1* uric acid transporter 1

the serum concentration of urate (Fig. 4.7). In addition, the 24 h urinary excretion of uric acid reflecting uric acid production is positively correlated with visceral and subcutaneous fat areas, suggesting that purine and fructose are excessively consumed in conjunction with overconsumption of calories and may contribute to an increase in the serum concentration of urate (Takahashi et al. 1997).

On the other hand, it is possible that hyperuricemia has an effect on adipocytes. A recent study demonstrated that uric acid produced reactive oxygen species (ROS) in adipocytes via NADPH oxidase activation, resulting in activation of MAP kinases p38 and ERK1/2, a decrease in nitric oxide bioavailability, and increases in protein nitrosylation and lipid oxidation. Oxidative stress and inflammation in adipose tissues induce imbalances in adipocyte-producing hormones and cytokines, which contribute to the development of insulin resistance and cardiovascular disease (Berg and Scherer 2005; Furukawa et al. 2004; Wellen and Hotamisligil 2005). Accordingly, hyperuricemia-induced ROS production by adipocytes may play a role in these disturbances. Metabolic syndrome increases the serum concentration of urate, leading to hyperuricemia, which induces ROS production in adipocytes. ROS produced by hyperuricemia may aggravate metabolic syndrome, resulting in the development of cardiovascular disease. However, further examinations are required to identify the prooxidant action of uric acid-producing ROS in humans.

4.3.3.2 Diabetes Mellitus

It is well known that DM causes cardiovascular disease and is closely related to obesity, which is thought to be an important risk factor for DM. Obesity, especially visceral fat obesity, causes insulin resistance, leading to type 2 DM with hyperinsulinemia (Hossain et al. 2007; Kahn and Flier 2000). Hyperinsulinemia enhances the tubular reabsorption of sodium via SMCT1 and that of uric acid via URAT1 in the kidneys, leading to hypertension and hyperuricemia, respectively

(Quiñones-Galvan and Ferrannini 1997; Strazzullo et al. 2006; Anzai et al. 2007). However, hyperglycemia-induced glucosuria causes a decrease in serum concentration of urate because glucosuria disturbs the reabsorption of uric acid (Moriwaki et al. 1995). Therefore, hyperuricemia is frequently observed in insulin-resistant DM, the so-called type 2 DM, without glucosuria. In contrast, it is not observed in insulin-deficient DM with glucosuria.

On the other hand, many studies have found that elevated urate level predicts the development of DM, suggesting that urate is an independent predictor of DM (Wang et al. 2011; Viazzi et al. 2011; Taniguchi et al. 2001; Nakanishi et al. 2003). Accordingly, uric acid may play a role in the development of DM. Recently, it was demonstrated that hyperuricemia was related to high blood sugar and high serum insulin levels in oxonate-treated rats. Administration of allopurinol (xanthine oxidase inhibitor) or benzbromarone (URAT1 inhibitor) decreased blood sugar and serum insulin levels, suggesting that uric acid contributes to elevated blood glucose and serum insulin levels (Nakagawa et al. 2006).

However, to determine whether uric acid is related to DM and metabolic syndrome, further experimental and clinical examinations are needed.

4.3.3.3 Chronic Kidney Disease

CKD is defined as kidney damage for 3 or more months consisting of structural or functional abnormalities, with or without decreased GFR, manifested by pathological abnormalities or abnormal levels of markers of kidney damage, including abnormalities in the composition of blood or urine or abnormalities revealed in imaging tests (Levey et al. 2003). Many studies have documented that CKD is associated with CVD, and affected patients are at higher risk for CVD than individuals in the general population (Sarnak et al. 2003; Foley et al. 1998; Elsayed et al. 2007). Cardiovascular risk factors postulated to be relevant to CKD are hypertension, dyslipidemia, DM, age, smoking, malnutrition, anemia, hyperhomocysteinemia, elevated fibrinogen, dysregulation of calcium and phosphate, oxidative stress, and inflammation, as well as male gender and advanced age (Yamamoto and Kon 2009). Patients with gout have a much higher prevalence of mild-to-moderate kidney disease as compared to the general population (Kutzing and Firestein 2008), while it has also shown that renal diseases are causative factors of hyperuricemia, with the most important causative factor being glomerular filtration rate, which reduces the renal excretion of uric acid (Talbott and Terplan 1960).

On the other hand, it has been suggested that hyperuricemia is a causative factor of renal disease (Johnson et al. 2005c), with several epidemiological prospective studies demonstrating that hyperuricemia is an independent risk factor of CKD (Iseki et al. 2004; Tomita et al. 2000). Representative results were noted in the Okinawa General Health Maintenance Association (OGHMA) screenings conducted in 1997 and 1999 (Iseki et al. 2001). Baseline data were collected from 6,403 individuals to examine the association of serum uric acid level with renal failure. The baseline level of serum urate was significantly higher in those in whom serum creatinine increased to greater than 1.4 mg/dl in men and 1.2 mg/dl in women (serum urate 6.8 ± 1.4 mg/dl) than in those in whom it did not (serum urate 5.8 ± 1.5 mg/dl) after 2 years. Furthermore, serum urate had a significant correlation with high serum creatinine. These findings suggest that baseline serum urate concentration is an independent predictor of early detection of renal failure. However, hyperuricemia-induced renal injury must be identified by use of a clinical intervention study with uric acid-lowering agents. Several clinical intervention studies have been performed and demonstrated that uric acid-lowering agents improve renal function together with lowering the serum concentration of urate (Siu et al. 2006; Goicoechea et al. 2010). In a randomized trial of 113 patients with estimated creatinine (eGFR) levels greater than 60 ml/min, serum urate levels were decreased in those treated with allopurinol (100 mg/day) from 7.8 ± 2.1 to 6.0 ± 1.2 mg/dl (means ± SDs) after 24 months, whereas serum urate levels did not change in the control subjects throughout the study period. In those allopurinol-treated subjects, eGFR increased by 1.3 ± 1.3 ml/min (means ± SDs) after 24 months, whereas that was decreased by 3.3 ± 1.2 ml/min in the controls. These findings suggest that allopurinol slows down the progression of renal diseases in patients with CKD, and serum urate may have a relationship to the progression of renal diseases (Goicoechea et al. 2010). However, those studies had small sample sizes, thus further clinical intervention studies using large numbers of subjects are required.

In recent experimental studies (Kang et al. 2002; Sánchez-Lozada et al. 2008), an increase in the serum concentration of urate developed afferent arteriopathy, glomerular hypertrophy, tubulointerstitial damage, and macrophage infiltration, together with increased glomerular pressure and hypertension. It has also been demonstrated that increases in MCP-1, renin, and COX-2 expressions by uric acid together with a decrease in NO by uric acid might contribute to uric acid-induced renal damage (Kang et al. 2002; Khosla et al. 2005; Cirillo et al. 2006) since MCP-1 is an important chemokine in the development of renal disease and hypertension, renin plays an important role in vasoconstriction, and COX-2 is a major regulator of renin expression (Kang et al. 2002; Harris and Breyer 2001). Furthermore, uric acid was found to increase production of MCP-1, fibronectin, and lysyl oxidase (LOX) in the proximal tubular cells (HK-2) (Yang et al. 2010). Fibronectin is an extracellular matrix biomarker that plays a role in extracellular matrix maturation, while LOX is also an extracellular biomarker that plays an important role in the oxidation of extracellular matrix, though LOX also participates in cell proliferation, intracellular signal responses, cell migration, and tissue development. Accordingly, uric acid is suggested to exert its profibrotic effects through upregulation of LOX and plays a role in the development and progression of hyperuricemia-induced renal injury.

Taken together, it is suggested that hyperuricemia aggravates CKD, thus contributes to cardiovascular disease via hyperuricemia-induced deterioration of CKD (Fig. 4.8).

4.3.3.4 Hypertension

Hypertension is an important cardiovascular risk factor and its prevalence has increased markedly over the past century, from a frequency of 6–11 % in the early 1900s to greater than 30 % today in the United States and from nearly zero to about 25–30 % in other countries, except for Europe and the United States (Johnson et al.

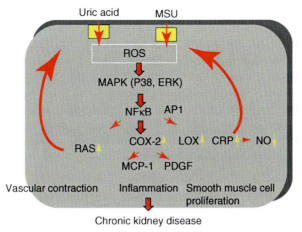

Fig. 4.8 Proposed mechanism of chronic kidney disease induced by uric acid and MSU. *LOX* lysyl oxidase, *ROS* reactive oxygen species, *MAPK* mitogen-activated protein kinase, *ERK* extracellular signal-regulated kinase, *NFκB* nuclear factor kappa-light-chain-enhancer of activated B cells, *AP-1* activator protein-1, *COX-2* cyclooxygenase-2, *RAS* renin-angiotensin system, *CRP* C-reactive protein, *NO* nitric oxide, *MCP-1* monocyte chemoattractant protein-1, *PDGF* platelet-derived growth factor

2005a). Similarly, the prevalence of hyperuricemia and gout has dramatically increased in Western countries as well as other countries. These increases in prevalence can be explained by lifestyle changes, such as food and alcohol consumption. In addition, the association between uric acid and hypertension has been recognized for many years, as 25–40 % of adults with untreated hypertension have hyperuricemia (6.5 mg/dl or higher) (Kinsey et al. 1961; Cannon et al. 1966). Conversely, individuals with hyperuricemia frequently have hypertension (60.7 %), as compared to those without hyperuricemia (30.5 %) (Schmidt et al. 1996). Although hyperuricemia is closely related to hypertension, it remains unresolved whether uric acid plays a direct role in the development of hypertension. Since hyperuricemia is also closely associated with renal function, obesity, and metabolic abnormalities, which profoundly affect hypertension (Vaccarino and Krumholz 1999), it is considered that hypertension develops from pathological conditions that elevated uric acid. In addition, it has also been proposed that hypertension directly decreases the urinary excretion of uric acid, leading to hyperuricemia, since a rise in blood pressure caused by infusion of noradrenaline or angiotensin was shown to decrease uric acid clearance (Yamamoto et al. 2001; Moriwaki et al. 2002).

On the other hand, recent studies have also suggested that uric acid is an independent risk factor for hypertension. For example, it was shown that uric acid levels predicted developing hypertension in men adjusting for age, body mass index, waist circumference, smoking, alcohol, plasma triglycerides, total cholesterol, and plasma glucose over a mean of 21.5 years of follow-up (Perlstein et al. 2006). Additionally, after adjusting for glomerular filtration rate in those subjects, uric acid levels remained significantly associated with hypertension over a mean 10.3 years of follow-up. In another study (Zhang et al. 2009), it was shown that uric acid levels independently predicted development of hypertension. In that prospective study, 7,220 participants without hypertension were followed for 4 years, of whom 1,370 men (19.0 %) and 208 women (11.0 %) developed hypertension. Uric acid levels predicted that development of hypertension in both sexes, after adjustment for risk factors related to hypertension (age, body mass index, smoking status, alcohol consumption, physical activity, blood glucose, triglycerides, total cholesterol, HDL cholesterol, creatinine, GFR, proteinuria, salt consumption, systolic and diastolic blood pressure, and family history of hypertension). Furthermore, after adjusting for abdominal obesity, as shown by waist circumference, uric acid levels remained significantly associated with developing hypertension. Understandably, patients with greater waist circumstance and those with metabolic syndrome have greater risk for developing hypertension than patients without higher waist circumstance and metabolic syndrome (Zhang et al. 2009). These results strongly suggest that renal function, metabolic syndrome, and abdominal obesity are involved in the pathogenesis of hypertension induced by uric acid. Recently, a meta-analysis regarding the longitudinal effect of uric acid on incident hypertension strongly suggested that hyperuricemia is a risk factor for developing hypertension (Grayson et al. 2011). In addition, in that study, female sex had a tendency to be associated with incident hypertension ($p=0.059$; Grayson et al. 2011). Thus, females with hyperuricemia have an increased risk for developing hypertension.

In several studies (Klein et al. 1973; Brand et al. 1985) uric acid more strongly predicted developing hypertension in younger than in elderly subjects. In addition, in cross-sectional epidemiological studies, the relationship of serum uric acid with blood pressure was stronger in younger individuals. Furthermore, a previous study found serum uric acid concentrations greater than 5.5 mg/dl in 89 % of the subjects with primary hypertension, while such elevation was not found in the controls (Feig and Johnson 2007). Uric acid is strongly and continuously correlated with systolic blood pressure and diastolic blood pressure ($r=0.81$, $p<0.001$; $r=0.66$, $p=0.001$, respectively), and these relationships cannot be explained by obesity or decreased renal function. A meta-analysis of prospective epidemiological studies showed that prevalence of hyperuricemia and hypertension were significantly and inversely related with mean study age (Grayson et al. 2011). Those epidemiological results strongly suggest that the impact of uric acid levels on blood pressure is stronger in the early stage of hypertension than in its late stage.

Interestingly, serum uric acid in adolescents with essential hypertension is inversely correlated with the birth weight of the affected individuals. In addition, subjects with primary hypertension were found to have lower birth weights as compared with control subjects and subjects with secondary or white coat hypertension (Feig and Johnson 2003). These results indicate that low birth weight predicts increased serum uric acid in adolescence, which may have a role in developing essential hypertension (Feig et al. 2004). In a pilot study (Feig et al. 2008), blood pressure in the new onset hypertensive adolescent subjects was shown to be significantly lowered by allopurinol treatment and then returned to the baseline level after washout of allopurinol. In addition, in a double-blind, placebo-controlled crossover study that included 30 hypertensive adolescents randomized to receive allopurinol or a placebo for 4 weeks, 86 % of the subjects in intervention group became normotensive following a decrease of uric acid below 5 mg/dl, while the corresponding rate of normotensive subjects was only 3 % in the control group (Feig et al. 2008). In another study (Kanbay et al. 2007), treatment with allopurinol significantly reduced blood pressure, CRP, and creatinine and increased GFR in asymptomatic hyperuricemic patients (mean age, 66.4 years) with normal renal functions. Another study found that systolic blood pressure was significantly elevated after allopurinol withdrawal in CKD patients without receiving renin-angiotensin system blockers, but not in those who received such blockers (Talaat and el-Sheikh 2007). Those findings suggest that uric acid induces hypertension in part via the renin-angiotensin system. The hypotensive effects observed in individuals receiving allopurinol treatment also suggest that uric acid has a casual role in the development of hypertension. However, since there is a possibility that the hypotensive effect can be attributed to the direct effect of allopurinol, such as xanthine oxidase inhibition but not to a hypouricemia effect, further studies are needed to investigate the effect of uric acid on hypertension development.

Recent experimental studies have shown that soluble uric acid has a direct effect on VSMCs (Kang et al. 205, Kanellis et al. 2003; Muraoka and Miura 2003; Mazzali et al. 2001). Soluble uric acid increases MCP-1 and PDGF in VSMCs, where URAT1 is localized (Kang et al. 2005b), resulting in VSMC proliferation (Kanellis et al. 2003). Those processes were mediated by activation of MAP kinases, MAP kinase-induced activation of nuclear transcription factor (NFκB) and activator protein-1 (AP-1), and stimulation of COX-2, while this uric acid-induced MCP-1 upregulation was inhibited by antioxidant substances, such as N-acetylcysteine and diphenyleneiodonium (Sautin et al. 2007; Kanellis et al. 2003; Muraoka and Miura 2003). Another report noted that soluble uric acid stimulated the proliferation and migration of human VSMCs and decreased NO release, in part, via CRP elevation. These effects of uric acid were reversed by probenecid, a URAT1 inhibitor, suggesting that vascular damage is induced by uric acid (Kang et al. 2005b). Therefore, VSMC damage induced by uric acid is mediated by increases of NADPH oxidase activity and ROS production, and a decrease in NO production. These uric acid effects were demonstrated *in vivo* in rats with mild hyperuricemia induced by a uricase inhibitor, which also showed elevated blood pressure and salt sensitivity.

Uric acid levels are correlated with blood pressure levels. Elevation of blood pressure can be prevented by treatment with either a xanthine oxidase inhibitor (allopurinol) or a uricosuric agent (benziodarone), both of which lower uric acid levels. Hyperuricemic rats exhibited an increase in juxtaglomerular renin and a decrease in macula densa neuronal NO synthase and afferent arteriolar thickening (Mazzali et al. 2001, 2002), and these findings strongly suggest that the renin-angiotensin system as well as NO reduction and salt sensitivity are involved in the pathogenesis of hypertension induced by uric acid. However, once preglomerular vascular disease develops, hypertension is driven by the kidney, and lowering uric acid levels is no longer protective (Watanabe et al. 2002). These results in animal models strongly support the notion that the impact of serum uric acid level on blood pressure is stronger in the early stage of hypertension.

4.3.3.5 Dyslipidemia

Hypertriglyceridemia is almost certainly an independent risk factor for CVD (Assmann et al. 1998; Murabito et al. 2012; Hokanson and Austin 1996) though the complex mechanism underlying the association between triglycerides and atherosclerosis obscures detection of their direct causal relationship. In a previous study, hypertriglyceridemia was reported to be frequently found in patients with hyperuricemia and gout (Takahashi et al. 1994). In addition, that study reported that the serum concentration of triglycerides was higher and that of HDL cholesterol was lower in patients with gout than in healthy subjects (Takahashi et al. 1994). Another study proposed a hypothetical mechanism regarding the relationship between hypertriglyceridemia and hyperuricemia as follows: insulin resistance complicated in gout patients reduces the activity of glyceraldehyde-3-phosphate dehydrogenase, resulting in accelerated triglyceride synthesis and purine synthesis (Leyva et al. 1998). However, hypertriglyceridemia is complicated in most of gout patients who have underexcretion of uric acid, suggesting that the relationship between hypertriglyceridemia and hyperuricemia is developed by another mechanism. Since insulin resistance reduces lipoprotein lipase (LPL) activity and increases triglyceride synthesis, decreased LPL activity and increased triglyceride synthesis may play roles in the increase in serum concentration of triglyceride in gout patients (Tsutsumi et al. 2001). Although the mechanism involved in the relationship between hypertriglyceridemia and hyperuricemia remains uncertain, hyperuricemia is suggested to develop insulin resistance, which increases the serum concentration of triglycerides. Accordingly, hyperuricemia may increase the serum concentration of triglycerides and contribute to cardiovascular disease via hypertriglyceridemia. Further examinations are needed to clarify the mechanism of hyperuricemia-induced hypertriglyceridemia in greater detail.

4.4 Summary

In this chapter, we described the causes of hyperuricemia, several of which cause CVD, as well as relationships of hyperuricemia with CVD, metabolic syndrome, diabetes mellitus, CKD, hypertension, and hypertriglyceridemia, which accelerate CVD, from epidemiological, clinical, and experimental findings. However, it remains uncertain whether hyperuricemia causes CVD. Accordingly, additional studies are needed to clarify causal relationship between hyperuricemia and CVD.

References

Abuja PM (1999) Ascorbate prevents prooxidant effects of urate in oxidation of human low density lipoprotein. FEBS Lett 446(2–3):305–308

Alderman MH, Cohen H, Madhavan S, Kivlighn S (1999) Serum uric acid and cardiovascular events in successfully treated hypertensive patients. Hypertension 34(1):144–150

Alonso A, Rodríguez LA, Logroscino G, Hernán MA (2007) Gout and risk of Parkinson disease: a prospective study. Neurology 69(17):1696–1700

Alper AB Jr, Chen W, Yau L, Srinivasan SR, Berenson GS, Hamm LL (2005) Childhood uric acid predicts adult blood pressure: the Bogalusa Heart Study. Hypertension 45(1):34–38

Alvarez-Lario B, Macarrón-Vicente J (2011) Is there anything good in uric acid? QJM 104(12):1015–1024

Ames BN, Cathcart R, Schwiers E, Hochstein P (1981) Uric acid provides an antioxidant defense in humans against oxidant- and radical-caused aging and cancer: a hypothesis. Proc Natl Acad Sci U S A 78(11):6858–6862

Anzai N, Kanai Y, Endou H (2007) New insights into renal transport of urate. Curr Opin Rheumatol 19(2):151–157

Assmann G, Cullen P, Schulte H (1998) The Münster Heart Study (PROCAM). Results of follow-up at 8 years. Eur Heart J 19(Suppl A):A2–A11

Beauvieux MC, Tissier P, Couzigou P, Gin H, Canioni P, Gallis JL (2002) Ethanol perfusion increases the yield of oxidative phosphorylation in isolated liver of fed rats. Biochim Biophys Acta 1570(2):135–140

Berg AH, Scherer PE (2005) Adipose tissue, inflammation, and cardiovascular disease. Circ Res 96(9):939–949

Bhole V, Choi JW, Kim SW, de Vera M, Choi H (2010) Serum uric acid levels and the risk of type 2 diabetes: a prospective study. Am J Med 123(10):957–961

Bos MJ, Koudstaal PJ, Hofman A, Witteman JC, Breteler MM (2006) Uric acid is a risk factor for myocardial infarction and stroke: the Rotterdam study. Stroke 37(6):1503–1507

Brand FN, McGee DL, Kannel WB, Stokes J 3rd, Castelli WP (1985) Hyperuricemia as a risk factor of coronary heart disease: the Framingham Study. Am J Epidemiol 121(1):11–18

Cannon PJ, Stason WB, Demartini FE, Sommers SC, Laragh JH (1966) Hyperuricemia in primary and renal hypertension. N Engl J Med 275(9):457–464

Chao HH, Liu JC, Lin JW, Chen CH, Wu CH, Cheng TH (2008) Uric acid stimulates endothelin-1 gene expression associated with NADPH oxidase in human aortic smooth muscle cells. Acta Pharmacol Sin 29(11):1301–1312

Chiou WK, Wang MH, Huang DH, Chiu HT, Lee YJ, Lin JD (2010) The relationship between serum uric acid level and metabolic syndrome: differences by sex and age in Taiwanese. J Epidemiol 20(3):219–224

Choi HK, Ford ES (2007) Prevalence of the metabolic syndrome in individuals with hyperuricemia. Am J Med 120(5):442–447

Choi HK, Liu S, Curhan G (2005) Intake of purine-rich foods, protein, and dairy products and relationship to serum levels of uric acid: the Third National Health and Nutrition Examination Survey. Arthritis Rheum 52(1):283–289

Choi HK, De Vera MA, Krishnan E (2008a) Gout and the risk of type 2 diabetes among men with a high cardiovascular risk profile. Rheumatology (Oxford) 47(10):1567–1570

Choi JW, Ford ES, Gao X, Choi HK (2008b) Sugar-sweetened soft drinks, diet soft drinks, and serum uric acid level: the Third National Health and Nutrition Examination Survey. Arthritis Rheum 59(1):109–116

Choi HK, Willett W, Curhan G (2010) Fructose-rich beverages and risk of gout in women. JAMA 304(20):2270–2278

Cirillo P, Sato W, Reungjui S, Heinig M, Gersch M, Sautin Y, Nakagawa T, Johnson RJ (2006) Uric acid, the metabolic syndrome, and renal disease. J Am Soc Nephrol 17(12 Suppl 3): S165–S168

Clifford AJ, Riumallo JA, Young VR, Scrimshaw NS (1976) Effect of oral purines on serum and urinary uric acid of normal, hyperuricemic and gouty humans. J Nutr 3:428–434

Constantinescu R, Zetterberg H (2011) Urate as a marker of development and progression in Parkinson's disease. Drugs Today (Barc) 47(5):369–380

Culleton BF, Larson MG, Kannel WB, Levy D (1999) Serum uric acid and risk for cardiovascular disease and death: the Framingham Heart Study. Ann Intern Med 131(1):7–13

Dehghan A, van Hoek M, Sijbrands EJ, Hofman A, Witteman JC (2008) High serum uric acid as a novel risk factor for type 2 diabetes. Diabetes Care 31(2):361–362

Dekker MJ, Su Q, Baker C, Rutledge AC, Adeli K (2010) Fructose: a highly lipogenic nutrient implicated in insulin resistance, hepatic steatosis, and the metabolic syndrome. Am J Physiol Endocrinol Metab 299(5):E685–E694

Di Chiara T, Argano C, Corrao S, Scaglione R, Licata G (2012) Hypoadiponectinemia: a link between visceral obesity and metabolic syndrome. J Nutr Metab 2012:175245

Dyer AR, Liu K, Walsh M, Kiefe C, Jacobs DR Jr, Bild DE (1999) Ten-year incidence of elevated blood pressure and its predictors: the CARDIA study. Coronary Artery Risk Development in (Young) Adults. J Hum Hypertens 13(1):13–21

Edwards NL (2008) The role of hyperuricemia and gout in kidney and cardiovascular disease. Cleve Clin J Med 75(Suppl 5):S13–S15

Elsayed EF, Tighiouart H, Griffith J, Kurth T, Levey AS, Salem D, Sarnak MJ, Weiner DE (2007) Cardiovascular disease and subsequent kidney disease. Arch Intern Med 167(11):1130–1136

Emmerson B (1998) Hyperlipidaemia in hyperuricaemia and gout. Ann Rheum Dis 57(9):509–510

Facchini F, Chen YD, Hollenbeck CB, Reaven GM (1991) Relationship between resistance to insulin-mediated glucose uptake, urinary uric acid clearance, and plasma uric acid concentration. JAMA 266(21):3008–3011

Faller J, Fox IH (1982) Ethanol-induced hyperuricemia: evidence for increased urate production by activation of adenine nucleotide turnover. N Engl J Med 307(26):1598–1602

Fang J, Alderman MH (2000) Serum uric acid and cardiovascular mortality the NHANES I epidemiologic follow-up study, 1971–1992. National Health and Nutrition Examination Survey. JAMA 283(18):2404–2410

Feig DI, Johnson RJ (2003) Hyperuricemia in childhood primary hypertension. Hypertension 42(3):247–252

Feig DI, Johnson RJ (2007) The role of uric acid in pediatric hypertension. J Ren Nutr 17(1):79–83

Feig DI, Nakagawa T, Karumanchi SA, Oliver WJ, Kang DH, Finch J, Johnson RJ (2004) Hypothesis: uric acid, nephron number, and the pathogenesis of essential hypertension. Kidney Int 66(1):281–287

Feig DI, Soletsky B, Johnson RJ (2008) Effect of allopurinol on blood pressure of adolescents with newly diagnosed essential hypertension: a randomized trial. JAMA 300(8):924–932

Foley RN, Parfrey PS, Sarnak MJ (1998) Epidemiology of cardiovascular disease in chronic renal disease. J Am Soc Nephrol 9(12 Suppl):S16–S23

Forman JP, Choi H, Curhan GC (2007) Plasma uric acid level and risk for incident hypertension among men. J Am Soc Nephrol 18(1):287–292

Freedman DS, Williamson DF, Gunter EW, Byers T (1995) Relation of serum uric acid to mortality and ischemic heart disease. The NHANES I Epidemiologic Follow-up Study. Am J Epidemiol 141(7):637–644

Furukawa S, Fujita T, Shimabukuro M, Iwaki M, Yamada Y, Nakajima Y, Nakayama O, Makishima M, Matsuda M, Shimomura I (2004) Increased oxidative stress in obesity and its impact on metabolic syndrome. J Clin Invest 114(12):1752–1761

Gao X, Qi L, Qiao N, Choi HK, Curhan G, Tucker KL, Ascherio A (2007) Intake of added sugar and sugar-sweetened drink and serum uric acid concentration in US men and women. Hypertension 50(2):306–312

Goicoechea M, de Vinuesa SG, Verdalles U, Ruiz-Caro C, Ampuero J, Rincón A, Arroyo D, Luño J (2010) Effect of allopurinol in chronic kidney disease progression and cardiovascular risk. Clin J Am Soc Nephrol 5(8):1388–1393

Grayson PC, Kim SY, LaValley M, Choi HK (2011) Hyperuricemia and incident hypertension: a systematic review and meta-analysis. Arthritis Care Res (Hoboken) 63(1):102–110

Greenway CV, Lautt WW (1990) Acute and chronic ethanol on hepatic oxygen ethanol and lactate metabolism in cats. Am J Physiol 258(3 Pt 1):G411–G418

Harris RC, Breyer MD (2001) Physiological regulation of cyclooxygenase-2 in the kidney. Am J Physiol Renal Physiol 281(1):F1–F11

Hein TW, Singh U, Vasquez-Vivar J, Devaraj S, Kuo L, Jialal I (2009) Human C-reactive protein induces endothelial dysfunction and uncoupling of eNOS in vivo. Atherosclerosis 206(1):61–68

Hokanson JE, Austin MA (1996) Plasma triglyceride level is a risk factor for cardiovascular disease independent of high-density lipoprotein cholesterol level: a meta-analysis of population-based prospective studies. J Cardiovasc Risk 3(2):213–219

Hossain P, Kawar B, El Nahas M (2007) Obesity and diabetes in the developing world – a growing challenge. N Engl J Med 356(3):213–215

Hunt SC, Stephenson SH, Hopkins PN, Williams RR (1991) Predictors of an increased risk of future hypertension in Utah. A screening analysis. Hypertension 17(6 Pt 2):969–976

Ichida K, Matsuo H, Takada T, Nakayama A, Murakami K, Shimizu T, Yamanashi Y, Kasuga H, Nakashima H, Nakamura T, Takada Y, Kawamura Y, Inoue H, Okada C, Utsumi Y, Ikebuchi Y, Ito K, Nakamura M, Shinohara Y, Hosoyamada M, Sakurai Y, Shinomiya N, Hosoya T, Suzuki H (2012) Decreased extra-renal urate excretion is a common cause of hyperuricemia. Nat Commun 3:764

Ikehara S, Iso H, Toyoshima H, Date C, Yamamoto A, Kikuchi S, Kondo T, Watanabe Y, Koizumi A, Wada Y, Inaba Y, Tamakoshi A, Japan Collaborative Cohort Study Group (2008) Alcohol consumption and mortality from stroke and coronary heart disease among Japanese men and women: the Japan collaborative cohort study. Stroke 39(11):2936–2942

Inokuchi T, Tsutsumi Z, Takahashi S, Ka T, Moriwaki Y, Yamamoto T (2010) Increased frequency of metabolic syndrome and its individual metabolic abnormalities in Japanese patients with primary gout. J Clin Rheumatol 16(3):109–112

Iseki K, Oshiro S, Tozawa M, Iseki C, Ikemiya Y, Takishita S (2001) Significance of hyperuricemia on the early detection of renal failure in a cohort of screened subjects. Hypertens Res 24(6):691–697

Iseki K, Ikemiya Y, Inoue T, Iseki C, Kinjo K, Takishita S (2004) Significance of hyperuricemia as a risk factor for developing ESRD in a screened cohort. Am J Kidney Dis 44(4):642–650

Johnson RJ, Titte S, Cade JR, Rideout BA, Oliver WJ (2005a) Uric acid, evolution and primitive cultures. Semin Nephrol 25(1):3–8

Johnson RJ, Feig DI, Herrera-Acosta J, Kang DH (2005b) Resurrection of uric acid as a causal risk factor in essential hypertension. Hypertension 45(1):18–20

Johnson RJ, Segal MS, Srinivas T, Ejaz A, Mu W, Roncal C, Sánchez-Lozada LG, Gersch M, Rodriguez-Iturbe B, Kang DH, Acosta JH (2005c) Essential hypertension, progressive renal disease, and uric acid: a pathogenetic link? J Am Soc Nephrol 16(7):1909–1919

Kahn BB, Flier JS (2000) Obesity and insulin resistance. J Clin Invest 106(4):473–481

Kanbay M, Ozkara A, Selcoki Y, Isik B, Turgut F, Bavbek N, Uz E, Akcay A, Yigitoglu R, Covic A (2007) Effect of treatment of hyperuricemia with allopurinol on blood pressure, creatinine clearence, and proteinuria in patients with normal renal functions. Int Urol Nephrol 39(4):1227–1233

Kanellis J, Watanabe S, Li JH, Kang DH, Li P, Nakagawa T, Wamsley A, Sheikh-Hamad D, Lan HY, Feng L, Johnson RJ (2003) Uric acid stimulates monocyte chemoattractant protein-1 production in vascular smooth muscle cells via mitogen-activated protein kinase and cyclooxygenase-2. Hypertension 41(6):1287–1293

Kang DH, Nakagawa T, Feng L, Watanabe S, Han L, Mazzali M, Truong L, Harris R, Johnson RJ (2002) A role for uric acid in the progression of renal disease. J Am Soc Nephrol 13(12):2888–2897

Kang DH, Park SK, Lee IK, Johnson RJ (2005a) Uric acid-induced C-reactive protein expression: implication on cell proliferation and nitric oxide production of human vascular cells. J Am Soc Nephrol 16(12):3553–3562

Kang DH, Han L, Ouyang X, Kahn AM, Kanellis J, Li P, Feng L, Nakagawa T, Watanabe S, Hosoyamada M, Endou H, Lipkowitz M, Abramson R, Mu W, Johnson RJ (2005b) Uric acid causes vascular smooth muscle cell proliferation by entering cells via a functional urate transporter. Am J Nephrol 25(5):425–433

Khosla UM, Zharikov S, Finch JL, Nakagawa T, Roncal C, Mu W, Krotova K, Block ER, Prabhakar S, Johnson RJ (2005) Hyperuricemia induces endothelial dysfunction. Kidney Int 67(5):1739–1742

Kim SY, Guevara JP, Kim KM, Choi HK, Heitjan DF, Albert DA (2010) Hyperuricemia and coronary heart disease: a systematic review and meta-analysis. Arthritis Care Res (Hoboken) 62(2):170–180

Kinsey DWR, Sise HS, Whitelaw G, Smithwick R (1961) Incidence of hyperuricemia in 400 hypertensive subjects. Circulation 24:972–973

Klein R, Klein BE, Cornoni JC, Maready J, Cassel JC, Tyroler HA (1973) Serum uric acid. Its relationship to coronary heart disease risk factors and cardiovascular disease, Evans County, Georgia. Arch Intern Med 132(3):401–410

Kobayashi S, Inoue N, Ohashi Y, Terashima M, Matsui K, Mori T, Fujita H, Awano K, Kobayashi K, Azumi H, Ejiri J, Hirata K, Kawashima S, Hayashi Y, Yokozaki H, Itoh H, Yokoyama M (2003) Interaction of oxidative stress and inflammatory response in coronary plaque instability: important role of C-reactive protein. Arterioscler Thromb Vasc Biol 23(8):1398–1404

Kolz M, Johnson T, Sanna S, Teumer A, Vitart V, Perola M, Mangino M, Albrecht E, Wallace C, Farrall M, Johansson A, Nyholt DR, Aulchenko Y, Beckmann JS, Bergmann S, Bochud M, Brown M, Campbell H, EUROSPAN Consortium, Connell J, Dominiczak A, Homuth G, Lamina C, McCarthy MI, ENGAGE Consortium, Meitinger T, Mooser V, Munroe P, Nauck M, Peden J, Prokisch H, Salo P, Salomaa V, Samani NJ, Schlessinger D, Uda M, Völker U, Waeber G, Waterworth D, Wang-Sattler R, Wright AF, Adamski J, Whitfield JB, Gyllensten U, Wilson JF, Rudan I, Pramstaller P, Watkins H, PROCARDIS Consortium, Doering A, Wichmann HE, Doering A, Wichmann HE, KORA Study, Spector TD, Peltonen L, Völzke H, Nagaraja R, Vollenweider P, Caulfield M, Spector TD, Peltonen L, Völzke H, Nagaraja R, Vollenweider P, Caulfield M, WTCCC, Illig T, Gieger C (2009) Meta-analysis of 28.141 individuals identifies common variants within five new loci that influence uric acid concentrations. PLoS Genet 5:e1000504

Kutzing MK, Firestein BL (2008) Altered uric acid levels and disease states. J Pharmacol Exp Ther 324:1–7

Kuzkaya N, Weissmann N, Harrison DG, Dikalov S (2005) Interactions of peroxynitrite with uric acid in the presence of ascorbate and thiols: implications for uncoupling endothelial nitric oxide synthase. Biochem Pharmacol 70(3):343–354

Lee HJ, Park HT, Cho GJ, Yi KW, Ahn KH, Shin JH, Kim T, Kim YT, Hur JY, Kim SH (2011) Relationship between uric acid and metabolic syndrome according to menopausal status. Gynecol Endocrinol 27(6):406–411

Levey AS, Coresh J, Balk E, Kausz AT, Levin A, Steffes MW, Hogg RJ, Perrone RD, Lau J, Eknoyan G, National Kidney Foundation (2003) National Kidney Foundation practice guidelines for chronic kidney disease: evaluation, classification, and stratification. Ann Intern Med 139(2):137–147

Leyva F, Wingrove CS, Godsland IF, Stevenson JC (1998) The glycolytic pathway to coronary heart disease: a hypothesis. Metabolism 47(6):657–662

Mäenpää PH, Raivio KO, Kekomäki MP (1968) Liver adenine nucleotides: fructose-induced depletion and its effect on protein synthesis. Science 161(847):1253–1254

Masson S, Desmoulin F, Sciaky M, Cozzone PJ (1992) The effects of ethanol concentration on glycero-3-phosphate accumulation in the perfused rat liver. A reassessment of ethanol-induced inhibition of glycolysis using 31P-NMR spectroscopy and HPLC. Eur J Biochem 205(1):187–194

Mazzali M, Hughes J, Kim YG, Jefferson JA, Kang DH, Gordon KL, Lan HY, Kivlighn S, Johnson RJ (2001) Elevated uric acid increases blood pressure in the rat by a novel crystal-independent mechanism. Hypertension 38(5):1101–1106

Mazzali M, Kanellis J, Han L, Feng L, Xia YY, Chen Q, Kang DH, Gordon KL, Watanabe S, Nakagawa T, Lan HY, Johnson RJ (2002) Hyperuricemia induces a primary renal arteriolopathy in rats by a blood pressure-independent mechanism. Am J Physiol Renal Physiol 282(6):F991–F997

Mendeloff AI, Weichselbaum TE (1953) Role of the human liver in the assimilation of intravenously administered fructose. Metabolism 2(5):450–458

Miller A, Adeli K (2008) Dietary fructose and the metabolic syndrome. Curr Opin Gastroenterol 24(2):204–209

Moriwaki Y, Yamamoto T, Takahashi S, Suda M, Higashino K (1995) Effect of glucose infusion on the renal transport of purine bases and oxypurinol. Nephron 69(4):424–427

Moriwaki Y, Yamamoto T, Tsutsumi Z, Takahashi S, Hada T (2002) Effects of angiotensin II infusion on renal excretion of purine bases and oxypurinol. Metabolism 51(7):893–895

Murabito JM, White CC, Kavousi M, Sun YY, Feitosa MF, Nambi V, Lamina C, Schillert A, Coassin S, Bis JC, Broer L, Crawford DC, Franceschini N, Frikke-Schmidt R, Haun M, Holewijn S, Huffman JE, Hwang SJ, Kiechl S, Kollerits B, Montasser ME, Nolte IM, Rudock ME, Senft A, Teumer A, van der Harst P, Vitart V, Waite LL, Wood AR, Wassel CL, Absher DM, Allison MA, Amin N, Arnold A, Asselbergs FW, Aulchenko Y, Bandinelli S, Barbalic M, Boban M, Brown-Gentry K, Couper DJ, Criqui MH, Dehghan A, den Heijer M, Dieplinger B, Ding J, Dörr M, Espinola-Klein C, Felix SB, Ferrucci L, Folsom AR, Fraedrich G, Gibson Q, Goodloe R, Gunjaca G, Haltmayer M, Heiss G, Hofman A, Kieback A, Kiemeney LA, Kolcic I, Kullo IJ, Kritchevsky SB, Lackner KJ, Li X, Lieb W, Lohman K, Meisinger C, Melzer D, Mohler ER 3rd, Mudnic I, Mueller T, Navis G, Oberhollenzer F, Olin JW, O'Connell J, O'Donnell CJ, Palmas W, Penninx BW, Petersmann A, Polasek O, Psaty BM, Rantner B, Rice K, Rivadeneira F, Rotter JI, Seldenrijk A, Stadler M, Summerer M, Tanaka T, Tybjærg-Hansen A, Uitterlinden AG, van Gilst WH, Vermeulen SH, Wild SH, Wild PS, Willeit J, Zeller T, Zemunik T, Zgaga L, Assimes TL, Blankenberg S, Boerwinkle E, Campbell H, Cooke JP, de Graaf J, Herrington D, Kardia SL, Mitchell BD, Murray A, Münzel T, Newman A, Oostra BA, Rudan I, Shuldiner AR, Snieder H, van Duijn CM, Völker U, Wright AF, Wichmann HE, Wilson JF, Witteman JC, Liu Y, Hayward C, Borecki IB, Ziegler A, North KE, Cupples LA, Kronenberg F (2012) Association between chromosome 9p21 variants and the ankle-brachial index identified by a meta-analysis of 21 genome-wide association studies. Circ Cardiovasc Genet 5(1):100–112

Muraoka S, Miura T (2003) Inhibition by uric acid of free radicals that damage biological molecules. Pharmacol Toxicol 93(6):284–289

Nakagawa T, Hu H, Zharikov S, Tuttle KR, Short RA, Glushakova O, Ouyang X, Feig DI, Block ER, Herrera-Acosta J, Patel JM, Johnson RJ (2006) A causal role for uric acid in fructose-induced metabolic syndrome. Am J Physiol Renal Physiol 290(3):F625–F631

Nakanishi N, Okamoto M, Yoshida H, Matsuo Y, Suzuki K, Tatara K (2003) Serum uric acid and risk for development of hypertension and impaired fasting glucose or type II diabetes in Japanese male office workers. Eur J Epidemiol 18(6):523–530

Niskanen LK, Laaksonen DE, Nyyssönen K, Alfthan G, Lakka HM, Lakka TA, Salonen JT (2004) Uric acid level as a risk factor for cardiovascular and all-cause mortality in middle-aged men: a prospective cohort study. Arch Intern Med 164(14):1546–1551

Perheentupa J, Raivio K (1967) Fructose-induced hyperuricaemia. Lancet 2(7515):528–531

Perlstein TS, Gumieniak O, Williams GH, Sparrow D, Vokonas PS, Gaziano M, Weiss ST, Litonjua AA (2006) Uric acid and the development of hypertension: the normative aging study. Hypertension 48(6):1031–1036

Price KL, Sautin YY, Long DA, Zhang L, Miyazaki H, Mu W, Endou H, Johnson RJ (2006) Human vascular smooth muscle cells express a urate transporter. J Am Soc Nephrol 17(7):1791–1795

Puig JG, Fox IH (1984) Ethanol-induced activation of adenine nucleotide turnover. Evidence for a role of acetate. J Clin Invest 74(3):936–941

Quiñones-Galvan A, Ferrannini E (1997) Renal effects of insulin in man. J Nephrol 10(4):188–191

Rao GN, Corson MA, Berk BC (1991) Uric acid stimulates vascular smooth muscle cell proliferation by increasing platelet-derived growth factor a-chain expression. J Biol Chem 266(13): 8604–8608

Richards J, Weinman EJ (1966) Uric acid and renal disease. J Nephrol 9:160–166

Ruidavets JB, Ducimetière P, Evans A, Montaye M, Haas B, Bingham A, Yarnell J, Amouyel P, Arveiler D, Kee F, Bongard V, Ferrières J (2010) Patterns of alcohol consumption and ischaemic heart disease in culturally divergent countries: the Prospective Epidemiological Study of Myocardial Infarction (PRIME). BMJ 341:c6077

Sánchez-Lozada LG, Tapia E, Soto V, Avila-Casado C, Franco M, Wessale JL, Zhao L, Johnson RJ (2008) Effect of febuxostat on the progression of renal disease in 5/6 nephrectomy rats with and without hyperuricemia. Nephron Physiol 108:69–78

Sarnak MJ, Levey AS, Schoolwerth AC, Coresh J, Culleton B, Hamm LL, McCullough PA, Kasiske BL, Kelepouris E, Klag MJ, Parfrey P, Pfeffer M, Raij L, Spinosa DJ, Wilson PW, American Heart Association Councils on Kidney in Cardiovascular Disease, High Blood Pressure Research, Clinical Cardiology, and Epidemiology and Prevention (2003) Kidney disease as a risk factor for development of cardiovascular disease: a statement from the American Heart Association Councils on Kidney in Cardiovascular Disease, High Blood Pressure Research, Clinical Cardiology, and Epidemiology and Prevention. Hypertension 42(5):1050–1065

Sautin YY, Nakagawa T, Zharikov S, Johnson RJ (2007) Adverse effects of the classic antioxidant uric acid in adipocytes: NADPH oxidase-mediated oxidative/nitrosative stress. Am J Physiol Cell Physiol 293(2):C584–C596

Schmidt MI, Watson RL, Duncan BB, Metcalf P, Brancati FL, Sharrett AR, Davis CE, Heiss G (1996) Clustering of dyslipidemia, hyperuricemia, diabetes, and hypertension and its association with fasting insulin and central and overall obesity in a general population. Atherosclerosis Risk in Communities Study Investigators. Metabolism 45(6):699–706

Siu YP, Leung KT, Tong MK, Kwan TH (2006) Use of allopurinol in slowing the progression of renal disease through its ability to lower serum uric acid level. Am J Kidney Dis 47(1):51–59

Strazzullo P, Barbato A, Galletti F, Barba G, Siani A, Iacone R, D'Elia L, Russo O, Versiero M, Farinaro E, Cappuccio FP (2006) Abnormalities of renal sodium handling in the metabolic syndrome. Results of the Olivetti Heart Study. J Hypertens 24(8):1633–1639

Sun SZ, Flickinger BD, Williamson-Hughes PS, Empie MW (2010) Lack of association between dietary fructose and hyperuricemia risk in adults. Nutr Metab (Lond) 7:16

Takahashi S, Yamamoto T, Moriwaki Y, Tsutsumi Z, Higashino K (1994) Impaired lipoprotein metabolism in patients with primary gout – influence of alcohol intake and body weight. Br J Rheumatol 33(8):731–734

Takahashi S, Yamamoto T, Tsutsumi Z, Moriwaki Y, Yamakita J, Higashino K (1997) Close correlation between visceral fat accumulation and uric acid metabolism in healthy men. Metabolism 46(10):1162–1165

Takahashi S, Moriwaki Y, Tsutsumi Z, Yamakita J, Yamamoto T, Hada T (2001) Increased visceral fat accumulation further aggravates the risks of insulin resistance in gout. Metabolism 50(4):393–398

Talaat KM, el-Sheikh AR (2007) The effect of mild hyperuricemia on urinary transforming growth factor beta and the progression of chronic kidney disease. Am J Nephrol 27(5):435–440

Talbott JH, Terplan KL (1960) The kidney in gout. Medicine (Baltimore) 39:405–467

Taniguchi Y, Hayashi T, Tsumura K, Endo G, Fujii S, Okada K (2001) Serum uric acid and the risk for hypertension and type 2 diabetes in Japanese men: the Osaka Health Survey. J Hypertens 19(7):1209–1215

Ter Maaten JC, Voorburg A, Heine RJ, Ter Wee PM, Donker AJ, Gans RO (1997) Renal handling of urate and sodium during acute physiological hyperinsulinaemia in healthy subjects. Clin Sci (Lond) 92(1):51–58

Tomita M, Mizuno S, Yamanaka H, Hosoda Y, Sakuma K, Matuoka Y, Odaka M, Yamaguchi M, Yosida H, Morisawa H, Murayama T (2000) Does hyperuricemia affect mortality? A prospective cohort study of Japanese male workers. J Epidemiol 10(6):403–409

Tsutsumi Z, Yamamoto T, Moriwaki Y, Takahashi S, Hada T (2001) Decreased activities of lipoprotein lipase and hepatic triglyceride lipase in patients with gout. Metabolism 50(8):952–954

Ugarte G, Iturriaga H (1976) Metabolic pathways of alcohol in the liver. Front Gastrointest Res 2:150–193

Vaccarino V, Krumholz HM (1999) Risk factors for cardiovascular disease: one down, many more to evaluate. Ann Intern Med 131(1):62–63

Viazzi F, Leoncini G, Vercelli M, Deferrari G, Pontremoli R (2011) Serum uric acid levels predict new-onset type 2 diabetes in hospitalized patients with primary hypertension: the MAGIC study. Diabetes Care 34(1):126–128

Wang T, Bi Y, Xu M, Huang Y, Xu Y, Li X, Wang W, Ning G (2011) Serum uric acid associates with the incidence of type 2 diabetes in a prospective cohort of middle-aged and elderly Chinese. Endocrine 40(1):109–116

Waslien CI, Calloway DH, Margen S (1968) Uric acid production of men fed graded amounts of egg protein and yeast nucleic acid. Am J Clin Nutr 21(9):892–897

Watanabe S, Kang DH, Feng L, Nakagawa T, Kanellis J, Lan H, Mazzali M, Johnson RJ (2002) Uric acid, hominoid evolution, and the pathogenesis of salt-sensitivity. Hypertension 40(3):355–360

Wellen KE, Hotamisligil GS (2005) Inflammation, stress, and diabetes. J Clin Invest 115(5):1111–1119

Wilk JB, Djousse L, Borecki I, Atwood LD, Hunt SC, Rich SS, Eckfeldt JH, Arnett DK, Rao DC, Myers RH (2000) Segregation analysis of serum uric acid in the NHLBI Family Heart Study. Hum Genet 106(3):355–359

Winkler K, Lundquist F, Tygstrup N (1969) The hepatic metabolism of ethanol in patients with cirrhosis of the liver. Scand J Clin Lab Invest 23(1):59–69

Woods HF, Eggleston LV, Krebs HA (1970) The cause of hepatic accumulation of fructose 1-phosphate on fructose loading. Biochem J 119(3):501–510

Yamamoto S, Kon V (2009) Mechanisms for increased cardiovascular disease in chronic kidney dysfunction. Curr Opin Nephrol Hypertens 18(3):181–188

Yamamoto T, Moriwaki Y, Takahashi S, Tsutsumi Z, Hada T (2001) Effect of norepinephrine on the urinary excretion of purine bases and oxypurinol. Metabolism 50(10):1230–1233

Yamamoto T, Moriwaki Y, Takahashi S, Tsutsumi Z, Ka T, Fukuchi M, Hada T (2002) Effect of beer on the plasma concentrations of uridine and purine bases. Metabolism 51(10):1317–1323

Yamamoto T, Moriwaki Y, Takahashi S (2005) Effect of ethanol on metabolism of purine bases (hypoxanthine, xanthine, and uric acid). Clin Chim Acta 356(1–2):35–57

Yang Z, Xiaohua W, Lei J, Ruoyun T, Mingxia X, Weichun H, Li F, Ping W, Junwei Y (2010) Uric acid increases fibronectin synthesis through upregulation of lysyl oxidase expression in rat renal tubular epithelial cells. Am J Physiol Renal Physiol 299(2):F336–F346

Yu MA, Sánchez-Lozada LG, Johnson RJ, Kang DH (2010) Oxidative stress with an activation of the renin-angiotensin system in human vascular endothelial cells as a novel mechanism of uric acid-induced endothelial dysfunction. J Hypertens 28(6):1234–1242

Zhang W, Sun K, Yang Y, Zhang H, Hu FB, Hui R (2009) Plasma uric acid and hypertension in a Chinese community: prospective study and metaanalysis. Clin Chem 55(11):2026–2034

Zharikov S, Krotova K, Hu H, Baylis C, Johnson RJ, Block ER, Patel J (2008) Uric acid decreases NO production and increases arginase activity in cultured pulmonary artery endothelial cells. Am J Physiol Cell Physiol 295(5):C1183–C1190

Proteomics Toward Biomarkers Discovery and Risk Assessment

Gloria Alvarez-Llamas, Fernando de la Cuesta, and Maria G. Barderas

Abstract

Cardiovascular diseases are among the leading causes of morbidity and mortality in Western societies and developing countries. The ability to investigate the complete proteome provides a critical tool toward elucidating the complex and multifactorial basis of cardiovascular biology, especially disease processes such as myocardial infarction, heart failure, stroke, and peripheral arterial disease. Different strategies have been used to discover novel potential biomarkers related to cardiovascular risk. It seems evident that a combination of biomarkers from different pathological pathways adds substantial prognostic information with respect to the risk of death from cardiovascular causes.

Keywords

Cardiovascular disease • Biomarkers • Risk assessment • Proteomics

5.1 Introduction

Cardiovascular science investigates and analyzes the biological systems involved in prevention, treatment, and development of disease (Van Eyk 2011). Since most acute presentations of cardiovascular diseases (CVD) (i.e., acute coronary

---

G. Alvarez-Llamas
Department of Immunology, IIS-Fundacion Jimenez Diaz, Madrid, Spain

F. de la Cuesta
Department of Vascular Physiopathology,
Hospital Nacional de Paraplejicos, SESCAM, Toledo, Spain

M.G. Barderas (✉)
Department of Vascular Physiopathology, Laboratorio de Fisiopatología Vascular,
Hospital Nacional de Paraplejicos, SESCAM, Edificio de Terapia 2ª planta,
Toledo 45071, Spain
e-mail: megonzalezb@sescam.jccm.es

syndrome, stroke, coronary heart disease, and congenital heart effects) bear a high mortality rate and severe complications, there has been a steady development of new treatments aimed to decrease them. However, a precise and rapid diagnosis is still mandatory to better select the most appropriate therapy in each clinical setting (Fig. 5.1). There are many approaches that aid to establish or rule out the correct diagnosis, and those based on novel biomarkers will constitute a powerful tool. In this way, there has been, and currently is, a need for technological development. In this context, proteomics arises as a potent implement involving strategies, instrumentation, and techniques to solve problems implicated in different research areas that are in continuous evolution. Both together (biomarkers and proteomics) notably augment the information obtained from traditional risk factors (hypertension, diabetes, hyperlipidemia, and smoking), illuminating novel disease mechanisms (Gerszten et al. 2011). Furthermore, it has been demonstrated that the combination of biomarkers from different pathological pathways adds substantial prognostic information with respect to the risk of death from cardiovascular causes (Zethelius et al. 2008). With all this in mind, we could associate the disease phenotype with individual proteins or protein profiles (Edwards et al. 2008; Vivanco et al. 2006;

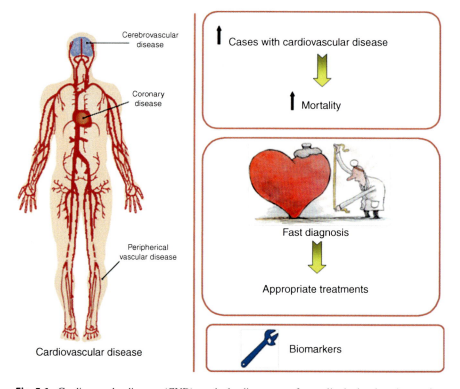

Fig. 5.1 Cardiovascular diseases (CVD) are the leading cause of mortality in developed countries. The development of new treatments is focused to decrease this situation. However, a precise and rapid diagnosis is the most appropriate tool in order to improve this situation

Arab et al. 2006) through the simultaneous analysis of a set of proteins which are present at a certain point of time in a particular cellular compartment, cell, tissue, or biological fluid. As a result, novel biomarkers of disease can be identified. One of the main problems in clinical practice is that the symptoms occur later in the course of the disease, and it is therefore mandatory to find out biomarkers of early diagnosis, prognosis, or evaluation of patient's recovery (Vivanco et al. 2011). However, before biomarkers are used clinically, it is important to elucidate important aspects of their clinical usefulness and applications, such as to establish their specific indications, to standardize the analytical methods to detect them, to assess their performance characteristics, and to study the incremental value and cost-effectiveness for given clinical indications (Alvarez-Llamas et al. 2008).

Whereas several drugs to treat risk factors such as dyslipidemia and hypertension have been developed over the past few decades, the identification of new biomarkers for early detection of cardiovascular disease has lagged behind (Kullo and Cooper 2010). Knowledge of new proteomic biomarkers may provide more precise estimation of risk while also defining the pathways perturbed in individual patients, revealing new targets for intervention and ultimately enabling an individualized approach to care (Cortese 2007).

5.2 Biomarkers

A general definition of a biomarker could be that established in 2001 by the NIH Biomarkers Definitions Working Group as follows: "a characteristic that is objectively measured and evaluated as an indicator of normal biological processes, pathogenic processes or pharmacological responses to therapeutic intervention" (Biomarkers Definitions Working Group 2001). According to this definition, a biomarker could either be a macromolecule, a measurement of a parameter (i.e., intima/media ratio), or an imaging procedure. However, a biomarker is widely considered to be a macromolecule, frequently a protein, whose levels are associated with a pathological process, with clinical value for diagnosis, prognosis, or therapeutic monitoring of a disease.

Clinical utility of molecular biomarkers relies on their specificity to predict pathological risk, together with their biological availability and measurement by inexpensive and easy implementable techniques (Fig. 5.2). In this sense, most biomarkers are blood proteins, since extraction and measurement of their levels can be performed with standardized commercial assays. Other fluids, like urine or cerebrospinal fluid (CSF), constitute upcoming sources for biomarker detection, since either they imply low-(non)invasive collection methods or they present less molecular complexity than blood plasma. Tissue biopsies constitute an interesting source of biomarkers for prognosis, but they are completely useless for early diagnosis purposes for obvious reasons.

Although many single proteins constitute biomarkers of pathology or a group of pathologies, there is a growing inclination to group altered proteins in order to create a biomarker panel of disease, which results in an increased specificity for outcome prediction compared to that of each of the proteins by themselves.

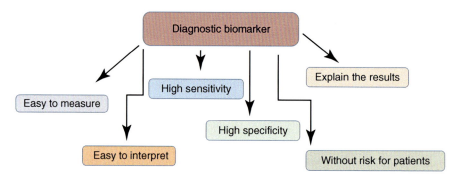

Fig. 5.2 Characteristics of a good biomarker

The acceptance of a biomarker by the scientific community implies an exhaustive validation in a great cohort of individuals, independent to those where the biomarker was discovered. Moreover, blind testing of the biomarker in such population has to be performed, comparing its ability to discriminate healthy and pathological subjects with respect to a well-established biomarker of the disease (Bossuyt et al. 2003), if available. Discrimination analysis is therefore the method of choice for new biomarker evaluation, and it is usually driven by the receiver operating characteristic (ROC) curve, a plot representing the true-negatives and false-positives assigned by the biomarker in the studied cohort (Alvarez-Llamas et al. 2008). The area under the curve (AUC), also called c-statistic, defines the ability of the biomarker to discriminate between diseased and healthy subjects (Manolio 2003). This parameter depends on the sensitivity and specificity of the test, where sensitivity refers to the ability to distinguish real pathological subjects from the healthy ones, the true-positive rate, and specificity is defined by the ability to exclude disease when it is not present, the true-negative rate (Gilstrap and Wang 2012). Maximum value of the AUC or c-statistic is 1, in which all the analyzed subjects are correctly classified as pathological or healthy, with no false-positives and false-negatives present. This ideal scenario is far away from the real situation, where c-statistic values of 0.7 or above are considered clinically relevant (Gilstrap and Wang 2012).

5.3 Development of New Biomarkers

In general, markers can be grouped into five categories according to their final purpose: (a) "risk assessment," grouping markers mainly responding to disease susceptibility; (b) "screening markers" which discriminate between healthy and asymptomatic disease in large populations; (c) "prognostic markers" able to predict probable course of disease or aggressiveness of therapy; (d) "stratification markers" focused to envisage responders and nonresponders to drug; and (e) "therapy monitoring," able to monitor the efficacy of treatment once the responder status is established (Finley Austin and Babiss 2006). One of the aims of the present chapter is to compile recent advances in the approaching way to discover new biomarkers of cardiovascular disease, which undoubtedly can be extended to any pathology of interest. Classical-biased approaches are intended to be placed in the context of an extended view, where multi-omic strategies are being developed and applied in a continuous increasing rate, being able to measure thousands of proteins or metabolites simultaneously.

The first question to be raised is "where to look for?" Depending on the disease, pathological mechanism, and drug action to be investigated, the starting material of choice is different. In general, the main technical challenges for discovery proteomics are associated with the existing hundreds or thousands of unique proteins and their potential posttranslational modifications (Schmidt et al. 1999). The overall dynamic range of individual protein levels that varies over 10 orders of magnitude in the body (Anderson and Anderson 2003; Kettman et al. 2002) is an important limitation if not correctly addressed (Kullo and Cooper 2010). In tissue, the rule is to try to discover relationships between cells and their products within functional compartments, and as close as possible to the foci of disease whenever possible (Marko-Varga and Fehniger 2004). Plasma (or serum) and urine are the most commonly used biological matrices in cardiovascular research, due to their availability and clinical relevance, as a source of potential biomarkers. Almost all cells in the body communicate with the plasma, either directly or through different tissues and biological fluids, releasing at least part of their intracellular content. By contrast, urine is produced by renal filtration of the plasma, and it is widely considered as one of the most important samples for diagnosis, as it does contain not only many plasma components but also the catabolic products of different metabolic pathways. Bronchoalveolar lavage, synovial fluid, saliva, amniotic, or cerebrospinal fluids are less common but can be easily related to a particular disorder and constitute more specific information sources. In this sense, these can be already considered as specific subproteomes compared to whole plasma or urine. They account with the advantage of tremendously diminishing protein abundance dynamic range, enormously facilitating the analysis from the technological side. Tissue secretome comprising the subset of proteins and peptides released by the different tissular compartments into the extracellular space constitutes an *ex vivo* approach which well simulates a zoom into the target organ activity and secretion to the plasma. This approximation allows detecting and investigating molecular pathways and dynamic responses of low-abundance molecules, which otherwise could not be detected in the whole plasma (de la Cuesta et al. 2012). Other strategies to reduce sample complexity and decrease limits of detection imply fractionation steps by liquid chromatography (ionic exchange or reversed phase) or fractionation based on isoelectric point differences (OFFGEL, Agilent Technologies; Rotofor, BioRad). In any case, the starting sample is divided in a considerable amount of subfractions with reduced complexity, increasing accordingly the number of identified proteins. However, time consuming should not be compromised in excess, keeping in mind that the easiest the technological platform and analytical methodology, the better in terms of translation to the clinical setting.

A step forward "in the biomarkers science" focuses on the whole set of proteins and/or metabolites (proteome and/or metabolome) without preselection of targets. The so generated data are not individual, referred to a unique molecule (protein), but global, describing hundreds or thousands of results simultaneously. This fact is due to the ability of the new techniques to characterize thousands of molecular species (proteins, lipids, metabolites, pharmaceuticals) in each run in a very short time frame, thus generating profiles or data sets which reflect the general situation of the sample (cell, tissue, biopsy, serum, urine, etc.). Thus, we are facing a new situation (termed "systems biology") where we deal with thousands of analytical data. With this concept in mind, different methodological approaches are currently available, making the choice mainly dependent on (a) the characteristics of the analytes to investigate (i.e., peptides, proteins, metabolites, lipids); (b) the performance offered by the technological platform in terms of sensitivity, selectivity, specificity, linear dynamic range, and throughput; and (c) the step in the biomarker research pipeline to approach (discovery or validation). In the discovery phase, gel-based platforms (2D-DIGE) and liquid chromatography (nLC-MS/MS) setups are most commonly used for protein analysis (Beer et al. 2011; Thakur et al. 2011), although the combination of capillary electrophoresis with mass spectrometry (CE-MS) for peptidome analysis is gaining popularity (Mischak et al. 2009). Metabolome differential analysis is currently approached by LC-MS, gas chromatography online coupled to mass spectrometry (GC-MS), and nuclear magnetic resonance (NMR) (Barderas et al. 2011; Rhee and Gerszten 2012). For those pathologies where spatial distribution analysis of proteins, peptides, and metabolites can be useful, mass spectrometry imaging (MSI) is the platform of choice (McDonnell and Heeren 2007; Wang et al. 2010). Once the potential biomarker candidate has been discovered, the next step of validation in a different cohort of samples should be approached. Apart from Western blot or ELISA, the analysis by selected reaction monitoring (SRM) is being increasingly established in current proteomics platforms. This strategy is typically performed in a triple quadrupole configuration MS instrument, being able to monitor and quantify, simultaneously, hundreds of molecules per sample through the measurement of specific fragments coming from the proteins/metabolites of interest (Lange et al. 2008).

It is not easy to imagine the complexity of interactions, which involve all of the biological activities and processes which are to be investigated. The main levels of study can be roughly divided in three components, function, expression, and structure (Mischak et al. 2009), but we need to include other components such as clinical dates (pathological or not), development stage, and state of the sample. Advances in proteomic technology, standardizing the collection and storage of specimens, preventing degradation during storage, and using well-defined cases and controls, will increase the likelihood of success in identifying novel markers that may help in disease prediction and prognostication. Last but not least, it is key to guarantee the access to bioinformatics' tools to analyze and interpret the output of proteomics studies in order to fully understand the biological mechanisms underneath. This challenge has also been described as the "Achilles heel of proteomics" by Patterson (2003).

At this point, it is worthwhile to point to the added difficulty to pave the way for translating the discovered biomarker into a routine clinical use, requiring the collaboration of the research laboratory, the diagnostics industry, and the clinical laboratory (Sturgeon et al. 2010). Are we ready for the biomarker industry? Once gaining the battle of technology and being able to reach the sensitivity demanded by the perfect specific candidates to be detected, the last issue is finding suitable application, proving clinical relevance, and gaining industry acceptance. It will be then that the benefit for patients, industry, and society has reached its maximum expression.

5.4 Cardiovascular Proteomics in Biomarker Discovery: A Proof of Concept

5.4.1 The Need for Early Diagnosis and Risk Assessment of CVD

Despite the raising advances in technology, an important number of cardiovascular diseases are diagnosed or treated when it is too late. At this time there are no solutions, or in best cases, surgery is the most indicated way to proceed. Predicting or detecting CVD in the earliest stage will translate into a higher cure rate.

This may be true but has to be proven because detecting CVD at the earliest stage may likewise lead to more intensive treatment with potential adverse treatment effects. In the case of CVD burden, there are some challenges to estimate changes in the incidence-prevalence of CVD. Although some epidemiologic studies have shown a downtrend in CV mortality, the trend in the incidence of CVD has demonstrated conflicting results (Roger et al. 2002). Some reasons for that are the change in diagnostic criteria for myocardial infarction, as well as in screening patterns, and the improved modalities to detect coronary and vascular disease in the subclinical phase, factors that will directly affect likelihood to diagnose a person with CVD (Batsis and Lopez-Jimenez 2010). Until recently, biomarker discovery efforts have focused on laborious approaches looking for that elusive singly overexpressed protein in blood (Petricoin and Liotta 2004), whose collection and processing are both simple and inexpensive (Veenstra et al. 2005). Furthermore, as blood circulates through every organ and tissue of the body, it contains valuable information pertaining to the physiological and pathological state of the organism (Anderson and Anderson 2002, 2003).

Nowadays, there is an increasing trend to avoid studying an exclusively biomarker for each CVD, but to identify a differential protein profile (biomarker panel). Thus, it seems reasonable that the presence of CVD and CV risk will be evaluated in the basis of multiplexed panels of clinical tests which will measure, at one time, the proteins increased or decreased as a consequence of CVD, increasing the discriminant power provided, when compared with that of the traditional risk factors alone. Using multiple protein biomarkers may lessen the impact of measurement error of any single biomarker alone, may be more sensitive and more specific, and may have the potential to improve risk stratification. Furthermore, protein biomarkers are attractive candidates to be included in risk assessment models due to their relative low cost, ease of attainment, and the information they provide about possible pathophysiological links to disease (Daniels 2011).

5.4.2 Traditional Cardiovascular Risk Factors and Their Use as Biomarkers of CVD

Although cardiovascular risk factors (hypertension, LDL cholesterol, aging, smoking) are biomarkers per se of cardiovascular diseases, they have been put together in an algorithm widely used by clinicians to calculate 10-year risk of having cardiovascular adverse outcomes, the Framingham Risk Score. This parameter was first established by Wilson et al. in 1998 (Wilson et al. 1998) and owes its name to the Framingham Heart Study, where over 2400 men and 2800 women were analyzed in terms of the aforementioned cardiovascular risk factors and risk stratified successfully with an algorithm adding or subtracting points in order to age, systolic blood pressure, LDL cholesterol, HDL cholesterol, and smoking habit, with a different value according to gender. Actually, the algorithm has been revised and for its calculation, clinicians follow the guidelines described by the Adult Treatment Panel III in 2002 (National Cholesterol Education Program (NCEP) Expert Panel on Detection, Evaluation, and Treatment of High Blood Cholesterol in Adults 2002). With these rules, Framingham Risk Score is calculated and, according to its value, stratification is built with three categories of increasing probability of having cardiovascular events in the next 10 years: low (<10 %), intermediate (10–20 %), and high risk (>20 %). An additional score derived from the Framingham Heart Study has recently arisen, which evaluates atrial fibrillation risk with promising results (Schnabel et al. 2009). The Framingham Risk Score has a c-statistic value for coronary heart disease (CHD) of 0.75 (Wilson 2011) which makes it to be considered by the scientific community as a good discriminant of the pathology. For this reason, finding new biomarkers with more discriminant capability than the Framingham Risk Score seems quite challenging, making more useful to combine them with this worldwide used parameter. Biomarkers may be considered clinically relevant if they increase the c-statistic of Framingham Risk Score in a valuable amount (at least 0.05) (Gilstrap and Wang 2012).

5.4.3 Heart and Vascular Protein Biomarkers in CVD: Translation to the Clinical Setting

The formation of the atheroma plaque is the causative factor that leads to coronary artery disease (myocardial infarction and angina), peripheral vascular disease (PAD), and cerebral vascular disease (ischemic stroke). In all cases, the process is asymptomatic so that the discovery of potential biomarkers is an urgent need. The scope is twofold: (1) further understanding of the operating mechanisms, for which tissue-based approaches are of great value (de la Cuesta et al. 2011), and (2) finding new targets for early diagnosis, on-time intervention, prognosis prediction, and therapy effect evaluation. To date, several soluble molecules are used by clinicians to predict future cardiovascular events, including C-reactive protein (CRP), B-type natriuretic peptides, and cardiac troponins (cTnI, cTnT). CRP is a protein released by the liver as a result of an inflammatory process which triggers interleukin 6

(IL-6) augment and, consequently, CRP stimulation by this cytokine (Libby and Ridker 2004). Its increase has been associated with cardiovascular events by many clinical trials (Halim et al. 2012). Although several acute-response proteins are expressed in the atherosclerotic plaque as a result of the internal inflammatory process which correlates with coronary risk (Libby and Ridker 2004), including serum amyloid A and CD40 ligand, besides IL-6 and CRP (Gilstrap and Wang 2012), the latter has arisen as the most useful inflammatory biomarker for cardiovascular risk prediction. Its utility as biomarker relies in its long half-life, ease of measurement, and the availability of inexpensive standardized high-sensitivity assays, as well as its high stability, not only due to its low thermosensibility but also due to the slight deviation of its levels in individuals by circadian variation or other external parameters (Libby and Ridker 2004). In the last years, high-sensitivity CRP (hsCRP) assays have been included in cardiovascular risk determination routine by cardiologists, since recent studies have demonstrated that its measurement adds prognostic information to the Framingham Risk Score, and it is a stronger predictor than parameters as important as plasma levels of LDL cholesterol (Ridker et al. 2002). Although its clinical utility has been widely proven, CRP levels are a consequence of an acute-phase response, and special care has to be taken with possible underlying infectious or inflammatory diseases that may produce false-positive results. Interestingly, recent studies have pointed out a specific role of CRP in ACS development due to prothrombotic capabilities of the monomeric protein dissociated from the pentameric circulating form by activated platelets (Molins et al. 2011).

The use of markers of atherosclerotic plaque instability as biomarkers for cardiovascular risk assessment is an interesting strategy widely explored in the last decades. These potential biomarkers expressed by unstable plaques have to be shed to the blood to have prognostic utility, which makes atheroma plaque secretome studies an ideal tool for this purpose (de la Cuesta et al. 2012). The development and vulnerability of atheroma plaque is a result of multiple molecular processes responding to lipid accumulation, inflammation, proteolysis, angiogenesis, hypoxia, apoptosis, thrombosis, and calcification. Among plaque instability biomarkers, lipoprotein-associated phospholipase A_2 (Lp-PLA$_2$) (Packard et al. 2000; Hatoum et al. 2011; Ballantyne et al. 2004) and myeloperoxidase (MPO) (Meuwese et al. 2007; Baldus et al. 2003; Mocatta et al. 2007; Morrow et al. 2008; Brennan et al. 2003) may be nowadays the ones with greater demonstrated clinical utility. A recent review based on carotid plaque compiles recent advances in plaque formation and serum biomarkers, to identify those at excess risk (Hermus et al. 2010). Most of them are mainly inflammatory and proteolytic markers such as hsCRP, SAA, IL-6, MMP-9, MMP-2, TIMP-1, and TIMP-2. There is certain controversy in the extended applicability of potential markers for coronary plaque instability to carotid artery disease, but considering that atherosclerosis is a systemic process sharing common risk factors and pathophysiological processes, related to plaque progression independently of its different location (coronary, carotid, or lower extremities), specificity of those biomarker candidates for a particular cardiovascular disorder cannot be guaranteed. In particular, PAD-specific biomarkers may exist in view of recent studies pointing to beta-2 microglobulin as a biomarker associated with PAD independently of other cardiovascular risk factors (Cooke and Wilson 2010).

Cardiac troponins cTnI and cTnT are considered the most robust biomarkers in detection of myocardial injury, thus acute myocardial infarction (AMI) diagnosis, as well as in risk stratification of ACS (Apple and Collinson 2012). Troponins are expressed by cardiac and skeletal muscles, and they are composed of three subunits: C, T, and I. The variants cTnI and cTnT, specific of the cardiac muscle (Hunkeler et al. 1991; Anderson et al. 1995), are shed to the blood after myocardial injury, which is frequently provoked by AMI (Gilstrap and Wang 2012). The detection of such troponins in the blood using high-sensitivity (hs) immunoassays allows clinicians to assess myocardial damage as well as to predict risk of future ACS (Reichlin et al. 2011). The use of cTnT was patented by Roche Diagnostics (Melanson et al. 2007), which owes the one high-sensitivity method available worldwide. Nevertheless, several hs-cTnI assays have been developed. Special care has to be taken with the epitopes reacting with the antibody, since cTnI is highly susceptible to proteolytic degradation, which is therefore less frequent in the central part of the molecule, where it locates cTnT interaction site, making epitopes in this zone more stable for such analysis (Waldo et al. 2008). Very recently, Zhang and co-workers analyzed chronic heart failure patients by top-down proteomics reporting a characteristic dephosphorylation of cTnI in Ser22 and Ser23, which makes these specific posttranslational modifications (PTMs) promising future biomarkers in heart failure risk assessment (Zhang et al. 2011). In addition, an upcoming proteomic approach with growing importance in macromolecules quantification such as SRM may be of great utility for high-sensitive determination of these biomarkers. Thus, Khun et al. described a method for cTnI quantification by SRM (Kuhn et al. 2009; Gerszten et al. 2010) which may be an interesting alternative to commercial immunoassays for implementation in medical facilities, which could benefit of the multiplexing measurement capabilities of a triple quadrupole for determination of several biomarkers in a single SRM analysis. Another useful protein for the diagnosis of AMI is heart-type fatty acid-binding protein (H-FABP) (Kleine et al. 1992), which has been recently shown to additionally predict stroke (Zimmermann-Ivol et al. 2004).

B-type natriuretic peptide (BNP) is a cardiac hormone secreted by the heart ventricles as a result of volume expansion and pressure overload (Waldo et al. 2008). B-type natriuretic peptides derive from preproBNP form, which is cleaved into a 108 amino acid sequence called proBNP (Levin et al. 1998). This protein is the precursor of the natriuretic peptides BNP, with vasodilatory activity, and NT-proBNP, which is the remaining peptide after proteolytic processing without known function, to date. The synthesis and cleavage of proBNP is triggered by pressure overload, and both BNP and NT-proBNP peptides are consequently released to plasma (van Kimmenade et al. 2006). Therefore, elevated levels of both B-type natriuretic peptides have been correlated by several clinical trials with heart failure (HF) (Di Angelantonio et al. 2009), making these peptides the prevalently selected biomarkers for HF diagnosis. Despite their utility, NT-proBNP is nowadays preferentially quantified due to its longer half-life. In addition, Hawkridge et al. reported in 2005 the absence of the peptide BNP within severe heart failure patients' blood when analyzing them with high-resolution LC-MS/MS, suggesting that altered forms of BNP are responsible for the signal obtained in less sequence-specific methods of this peptide's determination (Hawkridge et al. 2005). For this reason, NT-proBNP is a more adequate biomarker of HF than the BNP peptide. On the other hand, circulating proBNP has also been associated with acute heart failure, correlating with BNP and NT-proBNP blood levels, which indicates that this precursor protein may also be useful in heart failure diagnosis since its determination may exclude defects observed in natriuretic peptide levels due to body mass index (BMI) variation, making obesity a confounding variable (Waldo et al. 2008). Since BNP production has been directly associated with HF, recent studies point to an inducer of BNP production by angiotensin-II mediation, chromogranin B, as a new potential biomarker for the diagnosis of this pathology (Heidrich et al. 2008; Røsjø et al. 2010).

Despite the existence of these useful biomarkers of heart damage and heart failure, no early diagnosis biomarkers are available to date that may undoubtedly predict future events on healthy subjects. In this sense, proteomics has emerged as a powerful tool in the search for early diagnosis biomarkers, since it constitutes an undirected approach where the proteome of a sample is analyzed as a whole. Recent proteomic studies have pointed out potential biomarkers which may be implemented in clinical laboratories in the future, such as haptoglobin (Haas et al. 2011), apolipoprotein J (Cubedo et al. 2011), or leucine-rich α2-glycoprotein (Watson et al. 2011).

5.5 Predictive Utility

The burgeoning research in biomarker discovery requires a systematic organization of data with the use of standardized taxonomies that facilitates the online sharing of biomarker metadata among research (Vasan 2006). It is necessary to carry out a large number of epidemiological and clinical studies, screening quality, and cost-effectiveness of biomarkers in order to select those with excellent characteristics for helping patients.

The utility of individual cardiac biomarkers depends upon their ability to detect and stratify patients' risk with potential cardiovascular disease. In a hospital emergency department, the ideal cardiac biomarker will allow early detection of patients with a potential cardiovascular event and enable optimal treatment pathways to be rapidly initiated. Although their role in acute cardiovascular diseases, such as myocardial infarction and heart failure, has been well studied, no biological marker has emerged as the best screening marker for cardiovascular disease. Since a high number of patients arriving to the emergency department with potential cardiovascular disease do not ultimately have a cardiac etiology for their symptoms, a cardiac biomarker with a high negative predictive value would be useful to allow expeditious evaluation and discharge. Cardiac biomarkers with high positive predictive values are ideal to tailor aggressive care for patients at high risk of cardiovascular complications. To that end, we require a panel of cardiac biomarkers that provide nonoverlapping information to already existing (Wang 2011) and get a rapid "rule out" and identification of patients with elevated cardiovascular disease risk.

The current and future development of new proteomic tools will have important implications in risk assessment, therapeutic efficacy, diagnosis, etc. In a not too distant future, the cardiologist together with a specialist in proteomics, using MS as principal tool, could be able to earlier diagnose a particular cardiovascular event. Besides, an individualized selection of therapeutic combinations that best target the protein network will be available.

Several studies have tested whether multiple markers increase the ability to predict adverse cardiovascular outcomes (Wang 2006; Zethelius et al. 2008). These studies were able to improve the risk prediction by combining biomarkers into a multimarker score and the c-statistic for predicting myocardial infarction and death during a median follow-up of 10 years. Evaluation of diagnosis or predictive test uses the c-statistic as a measure of the test's ability to discriminate individuals with disease from those without diseases (Cook 2008). Because it is difficult to demonstrate improvements in the c-statistic, some investigators have advocated the use of other metrics to evaluate the predictive utility of new biomarkers (Cook 2007; Pencima et al. 2008).

However, in the area of proteomics, there is still a long way to go. The relationship among protein expression, regulation, and function within diseases has generated a global aim, compiled in the HUPO initiative (www.HUPO.org). Recent activities also addressed by the Human Proteome Project are to map and characterize human proteins in their biological context and develop innovative tools and reagents that the scientific community and, more specifically, the proteomics community can use to promote their understanding in the field and accelerate diagnostic, prognostic, therapeutic, and preventive medical applications.

There is still a long journey toward the identification of valuable clinical biomarkers and an arduous transition from the research environment to routine clinical practice. However, there is a clear need for harnessing emerging technologies in order to systematically assess variation in proteins with potential biomarker utility, which is unlikely to be found by focusing on well-studied pathways.

References

Alvarez-Llamas G, de la Cuesta F, Barderas ME, Darde V, Padial LR, Vivanco F (2008) Recent advances in atherosclerosis-based proteomics: new biomarkers and a future perspective. Expert Rev Proteomics 5(5):679–691

Anderson NL, Anderson NG (2002) The human plasma proteome: history, character, and diagnostic prospects. Mol Cell Proteomics 11:845–867

Anderson NL, Anderson NG (2003) The human plasma proteome: history, character, and diagnostic prospects. Mol Cell Proteomics 2(1):50

Anderson PA, Greig A, Mark TM, Malouf NN, Oakeley AE, Ungerleider RM, Allen PD, Kay BK (1995) Molecular basis of human cardiac troponin T isoforms expressed in the developing, adult, and failing heart. Circ Res 76(4):681–686

Apple FS, Collinson PO, IFCC Task Force on Clinical Applications of Cardiac Biomarkers (2012) Analytical characteristics of high-sensitivity cardiac troponin assays. Clin Chem 58(1):54–61

Arab S, Gramolini AO, Ping P, Kislinger T, Stanley B, Van Eyk J, Ouzounian M, MacLennan D, Emili A, Liu P (2006) Cardiovascular genomic medicine. Tools to develop novel biomarkers and potential applications. J Am Coll Cardiol 48:1733–1741

Baldus S, Heeschen C, Meinertz T, Zeiher AM, Eiserich JP, Münzel T, Simoons ML, Hamm CW, CAPTURE Investigators (2003) Myeloperoxidase serum levels predict risk in patients with acute coronary syndromes. Circulation 108:1440–1445

Ballantyne CM, Hoogeveen RC, Bang H, Coresh J, Folsom AR, Heiss G, Sharrett AR (2004) Lipoprotein-associated phospholipase A2, high-sensitivity C-reactive protein, and risk for incident coronary heart disease in middle-aged men and women in the Atherosclerosis Risk in Communities (ARIC) study. Circulation 109:837–842

Barderas MG, Laborde CM, Posada M, de la Cuesta F, Zubiri I, Vivanco F, Alvarez-Llamas G (2011) Metabolomic profiling for identification of novel potential biomarkers in cardiovascular diseases. J Biomed Biotechnol 2011:790132

Batsis JA, Lopez-Jimenez F (2010) Cardiovascular risk assessment-From individual risk prediction to estimation of global risk and change in risk in the population. BMC Med 8:29–34

Beer LA, Tang H, Barnhart KT, Speicher DW (2011) Plasma biomarker discovery using 3D protein profiling coupled with label-free quantitation. Methods Mol Biol 728:3–27

Biomarkers Definitions Working Group (2001) Biomarkers and surrogate endpoints: preferred definitions and conceptual framework. Clin Pharmacol Ther 69:89–95

Bossuyt PM, Reitsma JB, Bruns DE, Gatsonis CA, Glasziou PP, Irwig LM, Lijmer JG, Moher D, Rennie D, de Vet HC, STARD Group (2003) Towards complete and accurate reporting of studies of diagnostic accuracy: the STARD initiative. Clin Chem 49:1–6

Brennan ML, Penn MS, Van Lente F, Nambi V, Shishehbor MH, Aviles RJ, Goormastic M, Pepoy ML, McErlean ES, Topol EJ, Nissen SE, Hazen SL (2003) Prognostic value of myeloperoxidase in patients with chest pain. N Engl J Med 349:1595–1604

Cook NR (2007) Used and misuse of the receiver operating characteristic curve in risk prediction. Circulation 115:928–935

Cook NR (2008) Statistical evaluation of prognostic versus diagnostic models: beyond the ROC curve. Clin Chem 54:17–26

Cooke JP, Wilson AM (2010) Biomarkers of peripheral arterial disease. J Am Coll Cardiol 55:2017–2023

Cortese DA (2007) A vision of individualized medicine in the context of global health. Clin Pharmacol Ther 82:491–493

Cubedo J, Padró T, García-Moll X, Pintó X, Cinca J, Badimon L (2011) Proteomic signature of Apolipoprotein J in the early phase of new-onset myocardial infarction. J Proteome Res 10(1):211–220

Daniels LB (2011) Combining multiple biomarkers for cardiovascular risk assessment: more is usually better-up to a point. Bioanalysis 3(15):1679–1682

de la Cuesta F, Alvarez-Llamas G, Maroto AS, Donado A, Zubiri I, Posada M, Padial LR, Pinto AG, Barderas MG, Vivanco F (2011) A proteomic focus on the alterations occurring at the human atherosclerotic coronary intima. Mol Cell Proteomics 10(4):M110.003517

de la Cuesta F, Barderas MG, Calvo E, Zubiri I, Maroto AS, Darde VM, Martin-Rojas T, Gil-Dones F, Posada-Ayala M, Tejerina T, Lopez JA, Vivanco F, Alvarez-Llamas G (2012) Secretome analysis of atherosclerotic and non-atherosclerotic arteries reveals dynamic extracellular remodeling during pathogenesis. J Proteomics 75(10):2960–2971

Di Angelantonio E, Chowdhury R, Sarwar N, Ray KK, Gobin R, Saleheen D, Thompson A, Gudnason V, Sattar N, Danesh J (2009) B-type natriuretic peptides and cardiovascular risk: systematic review and meta-analysis of 40 prospective studies. Circulation 120:2177–2187

Edwards AVG, White MY, Cordwell SJ (2008) The role of proteomics in clinical cardiovascular biomarker discovery. Mol Cell Proteomics 7:1824–1837

Finley Austin MJ, Babiss L (2006) Commentary: where and how could biomarkers be used in 2016. AAPS J 8(1):E185–E189

Gerszten RE, Carr SA, Sabatine M (2010) Integration of proteomic-based tools for improved biomarkers of myocardial injury. Clin Chem 56(2):194–201

Gerszten RE, Asnani A, Carr SA (2011) Status and prospects for discovery and verification of new biomarkers of cardiovascular disease by proteomics. Circ Res 109:463–474

Gilstrap LG, Wang TJ (2012) Biomarkers and cardiovascular risk assessment for primary prevention: an update. Clin Chem 58(1):72–82

Haas B, Serchi T, Wagner DR, Gilson G, Planchon S, Renaut J, Hoffmann L, Bohn T, Devaux Y (2011) Proteomic analysis of plasma samples from patients with acute myocardial infarction identifies haptoglobin as a potential prognostic biomarker. J Proteomics 75(1):229–236

Halim SA, Newby LK, Ohman EM (2012) Biomarkers in cardiovascular clinical trials: past, present, future. Clin Chem 58(1):45–53

Hatoum IJ, Cook NR, Nelson JJ, Rexrode KM, Rimm EB (2011) Lipoprotein-associated phospholipase A2 activity improves risk discrimination of incident coronary heart disease among women. Am Heart J 161:516–522

Hawkridge AM, Heublein DM, Bergen HR 3rd, Cataliotti A, Burnett JC Jr, Muddiman DC (2005) Quantitative mass spectral evidence for the absence of circulating brain natriuretic peptide (BNP-32) in severe human heart failure. Proc Natl Acad Sci U S A 102(48):17442–17447

Heidrich FM, Zhang K, Estrada M, Huang Y, Giordano FJ, Ehrlich BE (2008) Chromogranin B regulates calcium signaling, nuclear factor kappaB activity, and brain natriuretic peptide production in cardiomyocytes. Circ Res 102(10):1230–1238

Hermus L, Lefrandt JD, Tio RA, Breek JC, Zeebregts CJ (2010) Carotid plaque formation and serum biomarkers. Atherosclerosis 213:21–29

Hunkeler NM, Kullman J, Murphy AM (1991) Troponin I isoform expression in human heart. Circ Res 69(5):1409–1414

Kettman JR, Coleclough C, Frey JR, Lefkovits I (2002) Clonal proteomics: one gene – family of proteins. Proteomics 2(6):624–631

Kleine AH, Glatz JF, Van Nieuwenhoven FA, Van der Vusse GJ (1992) Release of heart fatty acid-binding protein into plasma after acute myocardial infarction in man. Mol Cell Biochem 116(1–2):155–162

Kuhn E, Addona T, Keshishian H, Burgess M, Mani DR, Lee RT, Sabatine MS, Gersztein RE, Carr SA (2009) Developing multiplexed assays for troponin I and interleukin-33 in plasma by peptide immunoaffinity enrichment and targeted mass spectrometry. Clin Chem 55(6):1108–1117

Kullo IJ, Cooper LT (2010) Early identification of cardiovascular risk using genomics and proteomics. Nat Rev Cardiol 7(6):309–317

Lange V, Picotti P, Domon B, Aebersold R (2008) Selected reaction monitoring for quantitative proteomics: a tutorial. Mol Syst Biol 4:222

Levin ER, Gardner DG, Samson WK (1998) Natriuretic peptides. N Engl J Med 339:321–328

Libby P, Ridker PM (2004) Inflammation and atherosclerosis: role of C-reactive protein in risk assessment. Am J Med 116(Suppl 6A):9S–16S

Manolio T (2003) Novel risk markers and clinical practice. N Engl J Med 349(17):1587–1589

Marko-Varga G, Fehniger TE (2004) Proteomics and disease: the challenge for technology discovery. J Proteome Res 3:167–178

McDonnell LA, Heeren RMA (2007) Imaging mass spectrometry. Mass Spectrom Rev 26:606–643

Melanson SE, Tanasijevic MJ, Jarolim P (2007) Cardiac troponin assays: a view from the clinical chemistry laboratory. Circulation 116(18):e501–e504

Meuwese MC, Stroes ES, Hazen SL, van Miert JN, Kuivenhoven JA, Schaub RG, Wareham NJ, Luben R, Kastelein JJ, Khaw KT, Boekholdt SM (2007) Serum myeloperoxidase levels are associated with the future risk of coronary artery disease in apparently healthy individuals: the EPIC-Norfolk Prospective Population Study. J Am Coll Cardiol 50:159–165

Mischak H, Coon JJ, Novak J, Weissinger EM, Schanstra J, Dominiczak AF (2009) Capillary electrophoresis–mass spectrometry as a powerful tool in biomarker discovery and clinical diagnosis: an update of recent developments. Mass Spectrom Rev 28(5):703–724

Mocatta TJ, Pilbrow AP, Cameron VA, Senthilmohan R, Frampton CM, Richards AM, Winterbourn CC (2007) Plasma concentrations of myeloperoxidase predict mortality after myocardial infarction. J Am Coll Cardiol 49:1993–2000

Molins B, Peña E, de la Torre R, Badimon L (2011) Monomeric C-reactive protein is prothrombotic and dissociates from circulating pentameric C-reactive protein on adhered activated platelets under flow. Cardiovasc Res 92(2):328–337

Morrow DA, Sabatine MS, Brennan ML, de Lemos JA, Murphy SA, Ruff CT, Rifai N, Cannon CP, Hazen SL (2008) Concurrent evaluation of novel cardiac biomarkers in acute coronary syndrome: myeloperoxidase and soluble CD40 ligand and the risk of recurrent ischaemic events in TACTICS-TIMI 18. Eur Heart J 29:1096–1102

National Cholesterol Education Program (NCEP) Expert Panel on Detection, Evaluation, and Treatment of High Blood Cholesterol in Adults (Adult Treatment Panel III) (2002) Third Report of the National Cholesterol Education Program (NCEP) Expert Panel on Detection, Evaluation, and Treatment of High Blood Cholesterol in Adults (Adult Treatment Panel III) final report. Circulation 106(25):3143–3421

Packard CJ, O'Reilly DS, Caslake MJ, McMahon AD, Ford I, Cooney J, Macphee CH, Suckling KE, Krishna M, Wilkinson FE, Rumley A, Lowe GD (2000) Lipoprotein-associated phospholipase A2 as an independent predictor of coronary heart disease. West of Scotland Coronary Prevention Study Group. N Engl J Med 343:1148–1155

Patterson SD (2003) Data analysis – the Achilles heel of proteomics. Nat Biotechnol 21:221–222

Pencima NJ, D'Agostino RB Sr, D'Agostino RB Jr, Vasan RS (2008) Evaluating the added predictive ability of a new marker. From area under the ROC curve to reclassification and beyond. Stat Med 27:157–172

Petricoin EF, Liotta LA (2004) Proteomic approaches in cancer risk and response assessment. Trends Mol Med 10:59–63

Reichlin T, Irfan A, Twerenbold R, Reiter M, Hochholzer W, Burkhalter H, Bassetti S, Steuer S, Winkler K, Peter F, Meissner J, Haaf P, Potocki M, Drexler B, Osswald S, Mueller C (2011) Utility of absolute and relative changes in cardiac troponin concentrations in the early diagnosis of acute myocardial infarction. Circulation 124(2):136–145

Rhee EP, Gerszten RE (2012) Metabolomics and cardiovascular biomarker discovery. Clin Chem 58(1):139–147

Ridker PM, Rifai N, Rose L, Buring JE, Cook NR (2002) Comparison of C-reactive protein and low-density lipoprotein cholesterol levels in the prediction of first cardiovascular events. N Engl J Med 347(20):1557–1565

Roger LV, Jacobsen SJ, Weston SA, Goraya TY, Kilian J, Reeder GS, Kottke TE, Yawn BP, Frye RL (2002) Trends in incidence and survival of patients with hospitalized myocardial infarction, Olmsted County, Minnesota, 1979–1994. Ann Intern Med 136:341–348

Røsjø H, Husberg C, Dahl MB, Stridsberg M, Sjaastad I, Finsen AV, Carlson CR, Oie E, Omland T, Christensen G (2010) Chromogranin B in heart failure: a putative cardiac biomarker expressed in the failing myocardium. Circ Heart Fail 3(4):503–511

Schmidt AM, Yan SD, Wautier JL, Stern D (1999) Activation of receptor for advance glycation end products: a mechanism for chronic vascular dysfunction in diabetic vasculopathy and atherosclerosis. Circ Res 84:489–497

Schnabel RB, Sullivan LM, Levy D, Pencina MJ, Massaro JM, D'Agostino RB Sr, Newton-Cheh C, Yamamoto JF, Magnani JW, Tadros TM, Kannel WB, Wang TJ, Ellinor PT, Wolf PA, Vasan RS, Benjamin EJ (2009) Development of a risk score for atrial fibrillation (Framingham Heart Study): a community-based cohort study. Lancet 373(9665):739–745

Sturgeon C, Hill R, Hortin GL, Thompson D (2010) Taking a new biomarker into routine use – a perspective from the routine clinical biochemistry laboratory. Proteomics Clin Appl 4:892–903

Thakur SS, Geiger T, Chatterjee B, Bandilla P, Fröhlich F, Cox J, Mann M (2011) Deep and highly sensitive proteome coverage by LC-MS/MS without prefractionation. Mol Cell Proteomics 10(8):M110.003699

Van Eyk JE (2011) The maturing of proteomics in cardiovascular research. Circ Res 108:490–498

van Kimmenade RR, Januzzi JL Jr, Ellinor PT, Sharma UC, Bakker JA, Low AF, Martinez A, Crijns HJ, MacRae CA, Menheere PP, Pinto YM (2006) Utility of amino-terminal pro-brain natriuretic peptide, galectin-3, and apelin for the evaluation of patients with acute heart failure. J Am Coll Cardiol 48(6):1217–1224

Vasan RS (2006) Biomarkers of cardiovascular disease: molecular basis and practical considerations. Circulation 113:2335–2362

Veenstra TD, Conrads TP, Hood BL, Avellino AM, Ellenbogen RG, Morrison RS (2005) Biomarkers: mining the biofluid proteome. Mol Cell Proteomics 4:409–418

Vivanco F, Darde V, De la Cuesta F, Barderas MG (2006) Cardiovascular proteomics. Curr Proteomics 3:147–170

Vivanco F, De la Cuesta F, Barderas MG, Zubiri I, Alvarez-Lamas G (2011) Cardiovascular proteomics. In: Garcia A, Senis YA (eds) Platelet proteomics. Principles, analysis and applications. Wiley, Hoboken

Waldo SW, Beede J, Isakson S, Villard-Saussine S, Fareh J, Clopton P, Fitzgerald RL, Maisel AS (2008) Pro-B-type natriuretic peptide levels in acute decompensated heart failure. J Am Coll Cardiol 51(19):1874–1882

Wang TJ (2006) Multiple biomarkers for the prediction of first major cardiovascular events and death. N Engl J Med 355:2631–2639

Wang TJ (2011) Assessing the role of circulating, genetic, and imaging biomarkers in cardiovascular risk prediction. Circulation 123:551–565

Wang J, Balu N, Canton G, Yuan C (2010) Imaging biomarkers of cardiovascular disease. J Magn Reson Imaging 32(3):502–515

Watson CJ, Ledwidge MT, Phelan D, Collier P, Byrne JC, Dunn MJ, McDonald KM, Baugh JA (2011) Proteomic analysis of coronary sinus serum reveals leucine-rich α2-glycoprotein as a novel biomarker of ventricular dysfunction and heart failure. Circ Heart Fail 4(2):188–197

Wilson PW (2011) Prediction of cardiovascular disease events. Cardiol Clin 29:1–13

Wilson PW, D'Agostino RB, Levy D, Belanger AM, Silbershatz H, Kannel WB (1998) Prediction of coronary heart disease using risk factor categories. Circulation 97(18):1837–1847

Zethelius B, Berglund L, Sunström J, Ingelsson E, Basu S, Larson A, Venge P, Ärnlöv J (2008) Use of multiple biomarkers to improve the prediction of death from cardiovascular causes. N Engl J Med 358:2107–2116

Zhang J, Guy MJ, Norman HS, Chen YC, Xu Q, Dong X, Guner H, Wang S, Kohmoto T, Young KH, Moss RL, Ge Y (2011) Top-down quantitative proteomics identified phosphorylation of cardiac troponin I as a candidate biomarker for chronic heart failure. J Proteome Res 10(9):4054–4065

Zimmermann-Ivol CG, Burkhard PR, Le Floch-Rohr J, Allard L, Hochstrasser DF, Sanchez JC (2004) Fatty acid binding protein as a serum marker for the early diagnosis of stroke: a pilot study. Mol Cell Proteomics 3(1):66–72

Index

A

ABCG2 transporter, 89
ABC transporters, 70–74, 77, 83, 89–90
Acetaldehyde, 91, 92
Acetate, 91
Acetylcholine (ACh), 7–9, 14
Acetyl-CoA carboxylase, 49
Actin-myosin myofilament, 6, 11
Mitogen activated protein kinase (MAPK), 2, 11, 19–22, 26
Activator protein-1 (AP-1), 96–98, 103, 105
Acute coronary syndrome, 115–116
Acute myocardial infarction (AMI), 124
Adenylate cyclase (AC), 6–8
Adiponectin, 44, 46–48, 52
Adipose tissue, 43, 45–46, 48, 49, 56, 67–70, 88, 100
  epicardial adipose tissue (EAT), 51–52
  visceral adipose tissue (VAT), 48, 52
Adipose-tissue-myocardium axis, 48
Advanced glycation end products (AGEs), 45, 47, 54–55
Aequorin, 16
AGE-binding receptors (RAGEs), 55
Aging, 122
Alcohol, 58, 91, 93, 103, 104
Alcohol dehydrogenase (ADH), 91, 92
Aldehyde dehydrogenase (ALDH), 91, 92
Allopurinol, 101, 102, 104, 105
AMP-activated kinase (AMPK), 49
Amyloid A, 123
Angina, 122
Angiogenesis, 123
Angiotensin-converting enzyme (ACE), 97
Angiotensin II (AngII), 2, 5, 9, 11–14, 16–18, 20–22, 24, 55, 97, 103, 125
Angiotensinogen, 97
Antifibrinolysis, 21
Antihypertensive, 14, 23, 25–26
Antihypertensive therapy, 25
Apocynin, 96
Apolipoprotein J, 125
Apolipoproteins, 65, 125
  apoB, 65–69, 73–79, 81–83
  apoB-100, 68, 69, 73–77
  apoE, 65, 68, 73–78, 80
Arachidonic acid, 7, 9, 14, 17, 21
Area under the curve (AUC), 118
Arginine-vasopressin (AVP), 17, 18, 21
Arterial pressure, 3
Arterial wall diameter, 5
Arteriolar wall thickening, 13
Asymmetric dimethylarginine (ADMA), 5
Atheroma, 122, 123
Atherosclerosis, 10, 15, 41, 43, 44, 46–49, 52, 54, 55, 73, 75, 78, 79, 82, 83, 99, 123
AT1R, 17
AT2R, 6
Atrial fibrillation (AF), 41, 43, 122
Atypical PKC (aPKC), 18

B

Bariatric surgery, 56
Benzbromarone, 101
Benziodarone, 105
Bioinformatics, 120
Biological mechanisms, 120
Biomarkers, 81, 102, 115–126
Birth weight, 104
Blood pressure (BP), 2–26, 39, 41, 52, 56–58, 103–106, 122
Body fluid regulation, 24
Body fluid volume, 24
Body mass index (BMI), 40, 41, 48, 103, 104, 125
Bronchoalveolar lavage, 119
B-type natriuretic peptide (BNP), 122, 124–125

C

Ca²⁺-activated K⁺ channels, 9
Ca²⁺ channel blockers, 25
Ca²⁺ channels, 6, 11, 16–17, 25, 26, 53, 55
   receptor-operated (ROC), 16
   store-operated (SOC), 16, 17
   voltage-gated (VGCC), 16
Ca²⁺-dependent MLC phosphorylation, 16–18
Ca²⁺ entry, 6, 11, 16, 17
Ca²⁺-induced calcium release, 53
Ca²⁺ influx, 16–18, 22, 26
⁴⁵Ca²⁺ influx, 17
Calcium handling, 45, 53
Caldesmon (CAD), 11, 19, 22
Calmodulin, 11, 16
Calphostin C, 18, 19
Calponin (CAP), 11, 19, 21
cAMP production, 17
Capillary electrophoresis, 120
Cardiac muscle fibers, 3
Cardiac output (CO), 2–4, 24, 43
Cardiac remodeling, 3
Cardiac troponins (cTnI, cTnT), 122, 124
Cardiovascular biology, 115, 119, 125
Cardiovascular complications, 15, 125
Cardiovascular death, 57
Cardiovascular diseases (CVD), 3, 40–44, 47, 48, 52, 55–58, 78, 88, 91, 94–106, 115, 118–119, 121–122, 125
Cardiovascular events, 122, 123, 125, 126
Ca²⁺ release, 6, 11, 16–18
Carnitine palmitoyltransferase-1, 49
Carotid artery intima-media thickness (CIMT), 48
CD40, 123
Central obesity, 40
Ceramides, 44, 50, 51
Cerebral vascular disease, 122
Cerebrospinal fluids (CSF), 117, 119
c-fos, 21–22
cGMP, 6
Chelerythrine, 18, 20
Chemokines, 13
Cholesterol, 46, 55, 64–66, 68–73, 75–77, 80–83, 103, 104, 106, 122
Cholesterol balance, 71–73
Cholesteryl ester (CE), 64, 67–71, 73, 77
Cholesteryl ester exchange/transfer protein (CETP), 67, 70, 71, 74, 77, 80
Chronic kidney disease (CKD), 88, 94, 101, 103–106
Chronic vascular disease, 15
Chylomicrons, 64–69, 75–77, 81
Circulating AngII, 5, 10, 24

Classical PKC (cPKC), 18
Collagen, 22, 23
Collagen type I (CITP), 22, 23
Composition, 64–65, 82, 101
Congestive heart failure (CHF), 41–43
Connexin 40, 9
Constitutive COX1, 7
Coronary artery disease (CAD), 10, 41–48, 52, 54, 55, 122, 123
Coronary heart disease (CHD), 73, 79, 81–83, 96, 116, 122
Cortical preglomerular afferent arteriole, 16
Cost-effectiveness, 117, 125
CPI-17 protein, 19
C-reactive protein (CRP), 47, 48, 82, 98, 103, 105, 122–123
Creatinine, 101, 104, 105
c-Src, 21, 22
c-statistic, 118, 122, 126
Cyclic AMP/protein kinase A, 7–8
Cyclooxygenase-2 (COX-2), 97, 98, 102, 103, 105
Cyclooxygenases (COX), 6–8, 14
CYP4A2, 9
CYP4A14, 9
CYP4A2 inhibitors, 9
Cytochrome, 9
Cytokines, 19, 22, 123
Cytoskeletal proteins, 18

D

DAG. *See* Diacylglycerides (DAG); Diacylglycerol (DAG)
Dahl salt-sensitive rats, 12–14, 22
Deoxycorticosterone acetate (DOCA), 9, 12, 13, 24
Development stage, 120
Diabetes mellitus, 39–58, 79–81, 94, 100–101, 106, 116
Diabetic cardiomyopathy, 43–45, 51, 53, 54
Diacylglycerides (DAG), 44, 45, 47, 50, 51
Diacylglycerol (DAG), 6, 11, 18, 44, 45, 47, 50–51
Diacylglycerol acetyltransferase 1 (DGAT1), 50–51
Dietary sodium, 3
Dilated cardiomyopathy, 41
Diphenyleneiodonium, 105
DOCA-salt-treated SHR, 12
Dys-and Hyper-/Hypolipoproteinemias (D&HLP), 73–80, 82
Dyslipidemia, 41, 42, 46, 47, 80, 81, 94, 99, 101, 106, 117

Index

E
Early diagnosis, 117, 121, 122, 125
ECM inducer protein (EMMPRIN), 23
Ectopic fat, 46–47
EETs, 9–10
Electron transport chain, 53, 54
Endogenous tissue inhibitors of MMPs (TIMPs), 22–24
Endoperoxides, 5
Endoplasmic reticulum (ER), 6
Endothelial dysfunction, 2, 10, 14–15, 44, 46, 98
Endothelial lipase (EDL), 67, 74, 77
Endothelim-derived hyperpolarizing factor (EDHF), 2, 5, 6, 8, 9, 14
Endothelin-1 (ET-1), 2, 5, 9–14, 16, 20, 22, 26, 52, 96–98
Endothelin-1 (ET-1) clearance, 11, 12
Endothelin-converting enzyme (ECE), 11
Endothelium, 5–8, 11, 15, 43, 55, 98
Endothelium-dependent relaxation, 5–8, 14, 15
Endothelium-derived contracting factor (EDCF), 2, 5, 8, 10, 11
Endothelium-derived hyperpolarizing factor (EDHF), 2, 5, 6, 8, 9, 14
Endothelium-derived relaxing factors (EDRFs), 2, 5, 8, 13
Enzymes, 7, 11, 18, 24, 50–52, 56, 65–67, 71, 73, 76, 82, 88–89, 97, 98
Epicardial adipose tissue (EAT), 45, 51–52
Epidermal growth factor receptor (EGFR), 21, 22
ERK1/2 MAPK, 97, 100
17β-Estradiol, 17, 20
ETA/B receptor, 10
$ET_A R$, 6, 11–13
$ET_B R$, 6, 11–13
ETBR-deficient rats, 12
Ethanol, 91–93
Extracellular matrix (ECM), 5, 10, 21–24, 46, 55, 102

F
Familial combined, 73, 76–77
Familial juvenile hyperuricemic nephropathy, 89
Fatty acids (FAs), 45–47, 49, 50, 54, 57–58, 67
Fatty acid transport proteins (FATPs), 49
Ferret aorta, 17–19
Fibronectin, 102
Fibrotic changes, 22
Forearm blood flow, 7

Framingham Heart Study, 95–96, 122
Framingham Risk Score, 122, 123
Frank-Starling law, 3
Free cholesterol (FC), 64, 67–73, 79
Free fatty acids (FFAs), 44–52, 57, 66, 68–69, 71, 79, 80
Freeze-dried beer, 93
Fructokinase, 94
Fructose, 93–95, 100
Fructose-1-phosphate (F-1-P), 94, 95
Fura-2, 16

G
40Gap 27, 9
Gap junctions, 9
Genetically susceptible, 3
Genetic makeup, 3
Genome-wide association study (GWAS), 89
Glomerular capillary pressure, 25
Glomerular filtration rate (GFR), 13, 25, 101, 103
GLUT4, 49, 50, 52
Glyceraldehyde-3-phosphate (GAH-3-P) dehydrogenase, 106
Glycosylated hemoglobin, 42
Gö6976, 18
Gout, 87–89, 93–94, 99, 101, 103, 106
G protein-coupled, 7
Growth promoters, 5, 13, 14
Guanylate cyclase (GC), 5, 6

H
H-7, 19
Haptoglobin, 125
Heart failure, 10–11, 124–125
Heart-fatty acid binding protein (H-FABP), 124
Heart rate, 2–4
Heart-type fatty acid-binding protein (H-FABP), 124
Hepatic lipase (HL), 67, 69, 70, 76, 80
Hepato-steatosis, 79–80
12-HETE, 9, 10
15-HETE, 9
20-HETE, 9
High density lipoprotein (HDL), 64–74, 77, 79–82, 104, 106, 122
High sensitivity-CRP (hsCRP) assays, 123
Hind limb, 20
H_2O_2, 7
HOMA-IR, 99

Human umbilical vein endothelial cells (HUVECs), 96, 97
HUPO initiative, 126
Hypercholesterolemia, 5–6, 73–76
  primary, 75
  secondary, 76
Hyperglycemia, 41, 42, 46, 47, 54, 55, 57, 58
Hyperinsulinemia, 99, 100
Hyperleptinemia, 41
Hyperlipidemia, 73, 74, 76–77, 79–83, 116
Hyper-Lp(a), 74, 77–79
  primary, 78
  secondary, 78–79
Hypertension (HTN), 1–26, 41–48, 55, 81, 88, 94, 99–106, 116, 117, 122
Hypertriglyceridemia, 65, 73, 76, 81, 106
Hypertrophy, 4, 10, 12, 13, 18–19, 22, 23, 25–26
Hyperuricemia, 87–106
Hypoglycemia, 57, 58
Hypothalamus-pituitary-adrenal axis, 46–47
Hypoxanthine-guanine phosphoribosyl phosphorylase (HGPRT), 88, 89

I

Immunoblotting, 23
Inducible COX2, 7
Inflammation, 22, 42–44, 46–49, 51, 52, 54, 55, 81, 97, 98, 100, 101, 122–123
Insulin, 43, 47–51, 53–57, 79–80, 99–101
Insulin-receptor substrate 1 (IRS1), 47, 51, 55
Insulin resistance, 11, 41–44, 47–49, 51, 53, 54, 99–101, 106
Insulin resistance-related HTN, 11
Intercellular adhesion molecules (ICAM), 52
Interleukin-6 (IL-6), 46, 48, 52, 122–123
Intermediate density lipoprotein (IDL), 64–67, 69, 77, 81
Ischemic stroke, 122
1-(5-isoquinolinesulfonyl)-2-methylpiperazine (H-7), 18

J

Juxtamedullary efferent arteriole, 16

K

K^+ channels, 6, 9
KCl, 16, 17
2-Kidney 1-clip hypertensive rats, 12
K^+-induced contraction, 19
K^+ ions, 9

L

Lactate, 92, 93
L-arginase, 98
L-arginine, 5–7, 14
L-citrulline, 5
LC-MS/MS, 124–125
LDL cholesterol, 122, 123
LDL receptor related protein (LRP), 65, 68, 75–77
Lecithin-cholesterol acyltransferase (LCAT), 67, 71
Left ventricular compliance, 4
Left ventricular hypertrophy, 4, 23
Leptin, 44, 46–47, 56
Lesch-Nyhan disease, 88
Leucine-rich α2-glycoprotein, 125
Leukocyte adhesion, 13, 15
Lipase, 67, 68
Lipid accumulation, 123
12-Lipooxygenase, 10
12/15-Lipooxygenase, 9, 10
12/15-Lipooxygenase deficiency, 10
Lipoprotein(a) (Lp(a)), 61–66, 73, 74, 77–79, 81
Lipoprotein-associated phospholipase A_2 (Lp-PLA$_2$), 123
β-Lipoproteinemia, 74–77
Lipoprotein lipase (LPL), 65–70, 74, 76, 106
Lipoproteins (Lp), 64–83, 123
  high density lipoprotein (HDL), 64–74, 77, 79–82, 104, 106, 122
  intermediate density lipoprotein (IDL), 64–67, 69, 77, 81
  low density lipoprotein (LDL), 46, 52, 55, 62, 64–83, 122, 123
  Lp(a), 61–66, 73, 74, 77–79, 81
  receptors, 65, 68
  very low density lipoprotein (VLDL), 64–72, 74–78, 80, 81
Lipoprotein-X (LP-X), 79
L-NAME-treated rats, 12
Low-abundance molecules, 119
Low density lipoprotein (LDL), 46, 52, 55, 62, 64–83, 122, 123
Low-(non)invasive collection, 117
L-type Ca^{2+} channel, 53
Lysyl oxidase (LOX), 102, 103

M

Macrophage scavenger receptors, 48
Malonyl-CoA, 49
MAPK kinase (MEK), 11, 19, 20, 26
Mass spectrometry, 120
Mass spectrometry imaging (MSI), 120

Index

Matrix metalloproteinases (MMPs), 22–24, 82, 123
Mean arterial pressure (MAP), 3, 4
Mesangial cell hypertrophy, 10
Metabolic syndrome (MetS), 44, 51, 79–81, 83, 88, 94, 99–101, 104, 106
Metabolism, 18, 22–23, 44, 46, 48–52, 57, 58, 65–74, 78–80, 83, 88, 90–92, 94, 99, 103, 119–120
Microsomal ethanol-oxidizing system (MEOS), 91
Mitochondrial dysfunction, 45, 50, 53–54
Mitogen-activated protein kinases (MAPKs), 2, 11, 19–22, 26, 51, 96–98, 103
MLC phosphatase (MLCP), 11, 16, 19, 21
MMP-2, 22–24, 123
MMP-9, 22–24, 123
Monocytes chemoattractant protein-1 (MCP-1), 48, 52, 96–98, 102, 103, 105
Mononuclear cells, 13–14
Multifactorial disorder, 3
Myeloperoxidase (MPO), 123
Myocardial infarction (MI), 41–43, 48, 56–57, 73, 78, 83, 95, 121, 122, 125, 126
Myoendothelial gap junctions, 9
Myosin light chain (MLC) kinase (MLCK), 11

N

Na^+/Ca^{2+} exchanger, 16–18
N-acetylcysteine, 105
NADPH oxidase, 20, 22, 96, 97, 100, 105
Na^+/K^+ pump, 9
National Health and Nutrition Examination Survey (NHANES), 2, 40, 41, 95
Nephrosclerosis, 21
Nitric oxide (NO), 2, 5–8, 10, 13–15, 21, 22, 25, 43, 52, 98, 100, 102, 103, 105
Nitrotyrosine deposition, 13
Non-invasive collection. *See* Low (non) invasive collection
Norepinephrine, 8, 15, 16, 19, 25
NOS1 (nNOS), 5
NOS2 (iNOS), 5
NOS3 (eNOS), 5–7, 10, 21
NO synthase (NOS), 5–8, 14, 22, 25
Novel PKC (nPKC), 18
Nuclear transcription factor (NFκB), 96–98, 103, 105
Nω-monomethyl-L-arginine (L-NMMA), 7
Nω-nitro-L-arginine methyl ester (L-NAME), 7, 9, 12, 14, 22

O

Obesity, 3, 11, 39–58, 81, 99–100, 103, 104, 125
Omentin, 48
Outer medullary vasa recta, 16
Overweight, 40, 41, 43, 44, 47, 57
β-Oxidation, 49, 53, 54
Oxidative phosphorylation, 53, 54, 94
Oxidative stress, 7, 13, 21, 22, 45, 50, 51, 53–55, 57–58, 100, 101
Oxygen consumption, 4

P

PDBu, 19, 20
Peripheral arterial disease (PAD), 122–124
Peripheral resistance, 2, 25
Peripheral resistance vessels, 2, 25
Peripheral vascular disease, 122–124
Peritubular capillary, 13
Peroxisome proliferators-activated receptor α (PPAR-α), 49, 50
Peroxisome proliferators-activated receptor γ (PPAR-γ), 47, 79
Peroxynitrite (ONOO$^-$), 13
Phosphatidylinositol 4,5-bisphosphate (PIP$_2$), 6, 16, 18
Phospholipase A2, 7
Phospholipase C-β (PLCβ), 6, 11
Phospholipid, 18
Phospholipid exchange/transfer protein (PLTP), 67, 78
α-PKC, 18, 20–21
δ-PKC, 18, 20–21
ζ-PKC, 18–21
Plaque secretome, 123
Plasmalemmal Ca^{2+}-ATPase, 6, 17–18
Plasma volume, 2
Platelet-derived growth factor (PDGF), 96–98, 103, 105
p38 MAPK, 96, 97
P450 monooxygenase, 9
Porcine coronary artery, 18
p47 phox, 96
Predisposing factor, 3
Prehypertensive state, 25
Preload, 3, 24
PreproBNP, 124
Pressor response, 10
Pressure-natriuresis, 14, 24, 25
Probenecid, 96–98, 105
Procollagen type I amino-terminal propeptide (PINP), 23
Prognostic markers, 118, 120

Prostacyclin (PGI$_2$), 2, 5–9, 14
Prostaglandin synthases, 7
Prostanoids, 7, 8, 14
Protein kinase, 7–8, 11, 19, 21
Protein kinase C (PKC), 2, 11, 15, 17–22, 26, 51
Proteolysis, 123, 124
Proteomics, 54, 66, 83, 115–126
Proximal tubular cells, 102
Pseudosubstrate inhibitor peptide 19–36, 18
Purines, 90–91, 93, 100, 106

R
Raf, 11
Randle's glucose-fatty acid cycle, 49
Rat aorta, 8, 12, 18, 19, 24
Reactive oxygen species (ROS), 6, 7, 14, 21, 46, 47, 52–55, 82, 96–98, 100, 103, 105
Receiver operating characteristic (ROC), 118
Receiver operating characteristic (ROC) curve, 118
Receptor-operated Ca^{2+} channels (ROCs), 16
Renal autoregulatory adjustment, 25
Renal blood flow (RBF), 9, 25
Renal filtration, 119
Renal function, 2, 18, 26, 102–105
Renal microvessels, 13
Renin-angiotensin system (RAS), 13, 24, 26, 55, 97, 98, 103, 105
Resistance arteries, 5, 11, 12, 22, 24
Resistin, 48
Retinol-binding protein 4, 48
Reverse cholesterol transport (RCT), 65, 68, 70–71
RhoA, 21
Rho kinase, 2, 11, 15, 17, 21, 26
Rho/Rho-kinase signaling, 21
Risk assessment, 115–126
Ryanodine receptor, 53

S
SAA, 123
Saliva, 119
Salt intake, 25
Salt-loaded stroke-prone SHR, 12
Salt-sensitive HTN, 9, 14, 25
Sarcoplasmic reticulum (SR), 53
Sarcoplasmic reticulum Ca^{2+}-ATPase (SERCA), 6, 17–18
Secretome, 119, 123

Secretory type II phospholipase A$_2$ (sPLA$_2$-IIA), 52
Selected reaction monitoring (SRM), 120, 124
Skeletal muscles, 124
Small G protein, 21
Small resistance arteries, 5, 8
Smoking, 56, 58, 81, 101, 103, 104, 116, 122
Smooth muscle contraction, 2, 11
Sodium-calcium exchanger, 53, 55
Sodium-dependent monocarboxylate transporter 1 (SMCT1), 99, 100
Sodium-hydrogen exchanger, 55
Spontaneously hypertensive rat (SHR), 8–10, 12–15, 17–22, 24, 25
SR Ca^{2+} pump (SERCA2a), 53
Staurosporine, 18–20
Store-operated Ca^{2+} channels (SOCs), 16, 17
Stress, 3
Stress fibers, 21
Stroke, 3, 4, 12, 24, 43, 56–57, 73, 95, 115–116, 122, 124
Stroke volume, 2–4, 24
Subclinical phase, 121
Sudden cardiac death, 43
Superoxide, 52–54
Superoxide anion (•O$_2$–), 5, 7, 13, 20
Sympathetic nerve burst frequency, 3
Synovial fluid, 119
Systems biology, 83, 120
Systolic dysfunction, 42–45, 47, 50, 52

T
TAK-044, 10
Tangier disease (TD), 74, 77
Tempol, 97
TGF-β, 9–10, 22
Thromboxane A$_2$ (TXA$_2$), 2, 5, 8, 11, 13, 14, 16, 18, 21, 97
TIMP-1, 22–23, 123
TIMP-2, 123
Toll-like receptors (TLRs), 52
Total peripheral resistance (TPR), 3–5
α-Toxin, 19
Transcription factors, 21–22
Transplantation studies, 25
Triglycerides (TG), 45, 47, 50–51, 64–69, 71, 76, 77, 79–81, 103, 104, 106
Tubulointerstitium, 13
Tumor necrosis factor-α (TNF-α), 46, 48, 52
Type 2 diabetes mellitus (T2DM), 41, 44, 45, 47–52, 54–57, 94
Tyrosine kinase, 21, 22

Index

U

Uncoupling protein 1 (UCP-1), 51, 54
Uncoupling protein (UCP), 51, 54
Urate transporter 1 (URAT1), 89, 93, 96, 97, 99–101, 105
Uric acid, 87–91, 93–106
Urinary creatinine (UC), 11
Urinary 8-iso-PGF2α, 13
Urine, 117, 119, 120
Uromodulin, 89

V

Vascular dysfunction, 1–26, 43
Vascular endothelial cells (ECs), 5, 7, 14–15, 26
Vascular protection, 13, 17
Vascular reactivity, 12, 14, 15, 17, 19, 26
Vascular relaxation, 5–9, 15
Vascular remodeling, 2, 18, 22–24, 52
Vascular resistance, 2–5, 7, 13–15, 20, 24, 25
Vascular smooth muscle (VSM), 2, 5–22, 26, 43, 97
Vascular smooth muscle cells (VSMCs), 96, 97, 105
Vasoconstriction, 3, 5, 9–10, 12–17, 20–22, 25

Vasodilator receptors, 6, 8, 10
Very low density lipoprotein (VLDL), 64–70, 74–78, 80, 81
Visceral adipose tissue (VAT), 48, 52. *See* Adipose tissue
Visceral fat obesity, 99, 100
Visceral obesity, 40, 51
Visfatin, 48
Voltage-gated Ca^{2+} channels (VGCCs)
VSM hyperpolarization, 6, 9

W

Waist circumference, 40, 51, 103, 104
Water retention, 24
Weight loss, 55–56, 58
Weight-to-height ratio, 40
Wistar-Kyoto rat (WKY), 8, 15, 17, 19–21

X

Xanthine oxidase, 101, 105

Y

Y-27632, 21, 26